HUMAN VARIATION

A Genetic Perspective on Diversity, Race, and Medicine

HUMAN VARIATION

A Genetic Perspective on Diversity, Race, and Medicine

EDITED BY

Aravinda Chakravarti

McKusick–Nathans Institute of Genetic Medicine
Johns Hopkins University School of Medicine

COLD SPRING HARBOR LABORATORY PRESS
Cold Spring Harbor, New York • www.cshlpress.org

Human Variation: A Genetic Perspective on Diversity, Race, and Medicine

Chapters online at www.cshperspectives.org and www.perspectivesinmedicine.org

Executive Editor	Richard Sever
Managing Editor	Maria Smit
Senior Project Manager	Barbara Acosta
Permissions Administrator	Carol Brown
Production Editor	Diane Schubach
Production Manager	Denise Weiss
Cover Designer	Mike Albano
Publisher	John Inglis

Front cover artwork: A composite human face as a metaphor for the human genome, which can be thought of as a palimpsest containing recognizable genetic contributions from different ancestors and geographies. (Image created by Peter Jeffs.)

Library of Congress Cataloging-in-Publication Data

Human variation: a genetic perspective on diversity, race, and medicine/edited by Aravinda Chakravarti, Johns Hopkins University School of Medicine.
    page cm
"A subject collection from Cold Spring Harbor perspectives in medicine."
Includes bibliographical references and index.
ISBN 978-1-621820-90-1 (hardcover : alk. paper) -- ISBN 978-1-936113-25-5 (paper : alk.paper)
1. Human population genetics. 2. Human genetics--Variation. I. Chakravarti, Aravinda, editor of compilation.

QH455.H87 2013
599.93'5--dc23

                2012033597

10 9 8 7 6 5 4 3 2 1

All World Wide Web addresses are accurate to the best of our knowledge at the time of printing.

For a complete catalog of all Cold Spring Harbor Laboratory Press publications, visit our website at www.cshlpress.org.

Contents

Preface

WHAT INSANITY COMPELS A MAN TO CONTEMPLATE composing a book on genetics *and* race? To be sure, the content of this matter is more than science and an area of interest to many more than scholars in genetics. Moreover, one need not be a scholar of anything to have an opinion on both subjects, and, indeed, people all over the world have strong opinions on genetics and race, one way or another. We all know of such people among our relatives, neighbors, and colleagues. In one way, this is a distinct advantage because it assures us of a large audience for this book; it is also a planned disadvantage because it invites immediate and vociferous criticism from a great diversity of experts and nonexperts alike. No doubt much ink has been spilled on this topic, so why try again? My reason for this book is simply that the science of genetics has changed in the past 10 to 20 years in such a major and fundamental way that we need to recount and explain what we understand, *and do not know*, of human history, diversity, and race from this new vantage point. My reason for asking some of the noted scholars of today to do the majority of this retelling in this volume is not to deflect blame, or to dilute criticism directed at me, but to have real experts tell their own stories that have appeared through their own research. My role has been one of synthesis and editing.

This book, unfortunately, has had a very long gestation. But the idea of this retelling goes much further back. As a graduate student, I was one of very few in the 1970s who was interested and trained in the triple areas of human genetics, molecular biology, and human disease biology, largely as a result of some farsighted mentors. From this perspective, there was no way to avoid questions of genes and race, an age-old topic in the United States. Like many, I have wrestled with the same question that exists till today: What does genetics have to say about human differences? Attempts to answer this very broad query, in the context of human disease, have been a part and parcel of my academic life and research. Fortunately, for me, these attempts at understanding have fundamentally changed my ideas and views on this topic over time. It is not merely one more fossil, one more piece of the genetic puzzle, or the genetic features of one more disease that has changed my previously deeply held views. It is, rather, that my that knowledge of genes and genomes, how they function, and how they have led to a revision of the manner and timescale of human evolution and the intensity of natural selection and adaptation has changed in modern genetics. In turn, this has had a major impact on my thinking about how human diversity impacts the genetic architecture of human trait and disease differences. I am sure I am not alone in this discovery nor in changing my views. This is a second reason for me to recruit other scientists to tell of this transformation.

Our cumulative knowledge of human prehistory has been altered by advances in the science of genetics and contemporary studies of human diversity. Our cumulative knowledge of the nature of genetic inheritance of complex traits, natural selection, and adaptation has also changed. Finally, the nature of society in which the old debates of race and genes took place has also changed, for better or for worse. Scientists who have contributed to this store of knowledge deserve to tell this story in their own words. We have, for a long time, relied mostly on reporters and commentators outside the academy to tell this story on our behalf. It is about time that we did so ourselves, with greater clarity and with more nuance as our science dictates. Ignoring scientific details for the beauty of storytelling is precisely why genetics is often in hot water over questions about human diversity.

This book contains contributions by geneticists, biomedical scientists, social scientists, and a historian. I thank each of them for their care in exposition and for their patience during multiple revisions. Each has given their perspective on human diversity and history, with some speaking directly to the genetic bases of human trait and disease differences where they are understood. My role as an editor has not been to vet their views but to ensure that these essays are eminently readable by nongeneticists without sacrificing scientific rigor or details. The book could have been longer, and surely taken much longer, but the essential story of human diversity and history and the genetic meanings of race or ethnicity or any other human division are all there. It is clear today that humans differ much more in their languages, cultures, and societies than they do in their genes. This does not mean that the effects of genes can be sidestepped or ignored: quite the contrary, they can be invaluable tools to help us to understand the broad causes of human disease.

The reality of this publication is almost all due to my coauthors and the editorial staff at Cold Spring Harbor Laboratory Press. I wish to particularly thank Barbara Acosta for her extreme patience and great disposition and Richard Sever for being indulgent to my calendar. I hope that you find the book and its essays thought-provoking and fresh. The science is no longer controversial but its implications are astounding: despite our genetic diversity, we are truly one human race.

ARAVINDA CHAKRAVARTI, PH.D.
Baltimore, Maryland
July 2014

Perspectives on Human Variation through the Lens of Diversity and Race

Aravinda Chakravarti

McKusick-Nathans Institute of Genetic Medicine, Johns Hopkins University School of Medicine, Baltimore, Maryland 21205

Correspondence: aravinda@jhmi.edu

Human populations, however defined, differ in the distribution and frequency of traits they display and diseases to which individuals are susceptible. These need to be understood with respect to three recent advances. First, these differences are multicausal and a result of not only genetic but also epigenetic and environmental factors. Second, the actions of genes, although crucial, turn out to be quite dynamic and modifiable, which contrasts with the classical view that they are inflexible machines. Third, the diverse human populations across the globe have spent too little time apart from our common origin 50,000 years ago to have developed many individually adapted traits. Human trait and disease differences by continental ancestry are thus as much the result of nongenetic as genetic forces.

Half-jokingly, Gwen Ifill, the noted American journalist and newscaster, told the Smithsonian audience, "In no universe is President Obama white!" (Smithsonian's National Museum of Natural History 2013). Her comment came in response to my genetic argument that given the President's white American mother and Kenyan father, he could just as well be called "white" as "black."

This friendly exchange exposed the essential conundrum surrounding the contemporary meaning of labels, classification, and the notion of race. We humans have, since time immemorial, sorted and classified each other into numerous categories based on language, culture, and appearance (en.wikipedia.org/wiki/Race_(human_classification); Blumenbach 1775). Irrespective of how these groupings were decided on or justified, such classification has been a cultural exercise: the basis for self-identification through the identification of others. Genetics, being a modern science, has come to this scene much later. Genetics has much to say about the recent and remote history of our species and our individual ancestries, as well as the potential to support or refute our existing classifications. What genetics says about our history cannot be wished away. At the same time, our cultures have a strong voice in how we view ourselves and view others. This cultural view cannot be wished away either. President Obama's self-identification as "black" is not based on his personal gene accounting but rather a nonchoice given American social convention (the "one-drop" rule) and his personal history. That was Ifill's point. What we, and others, call us depends on both our genes and our society. I am a Bengali-American, the duality and hyphen being equally important to my identity over and above my genes.

So, what are we to think of the existence of human "races" in this "genomic age" and why is there still so much controversy? (Koenig et al. 2008). By any account, we cannot ignore the idea of human races. It is in daily common use, a basis of self-identification and for many a key to their social identity. Without quibbling about word usage and specific meaning, race is also the basis for governmental statistical accounting and political action. Although the word race finds daily use in the United States, and, increasingly in Europe, similar controversies surround other classifications of humans, such as caste or tribe, despite the diverse origins of each term (Thapar 2014). The most controversial aspect of such classifications, in my opinion, is not whether they are biological or not, but rather the imputation of wholesale traits[1] and attributes to these groups so defined. Almost without exception, the characteristics displayed by one's own group are deemed positive and implicitly valued, whereas those of other groups are deemed negative and are undervalued. The group defining and possessing the valued features is also invariably that group that is culturally dominant and politically powerful.

Genetics has always been an ideal fuel for this fodder. It is a science that examines the biological basis for trait differences, and has recently allowed us to make tremendous strides in our understanding of human disease (e.g., why some individuals have muscular dystrophy and others do not) and differences (e.g., why some individuals can digest milk and others cannot). The argument goes that if groups can differ in traits such as lactose intolerance (most Europeans and some Africans are tolerant, whereas others of the world are not) and malarial susceptibility (many Africans, Middle Easterners, South Asians, and East Asians have some protection, whereas the rest of the world does not) owing to specific gene differences, they, in all probability, also geneti-

cally differ in a host of health-related traits. But, why should this principle be restricted to health-related features? Some have argued, why not genetic differences in intellectual ability, industriousness, the facility for democratic institutions, or aggressive behavior (Wade 2014)? In the past 100 years of genetics, many in the field have advertently and inadvertently engaged in considerable speculation as to these last possibilities, the biological underpinnings of *any* difference. But what is the evidence that these metafeatures are genetic? I know of no science that can prove the genetic underpinnings of these broad social differences. In contrast, I know of plenty of evidence that argues against it being the case (Chakravarti 2010). As the physicist Neil deGrasse Tyson recently quipped: "You get to say the world is flat because we live in a country that guarantees free speech, but it is not a country that guarantees that anything you say is correct" (deGrasse Tyson 2014).

Over the next several pages, I would like to tell you what we know of human diversity and how we came to be who we are today: in short, our singular genetic heritage and history (Fig. 1). I will outline what we know today about genetic differences across contemporary human populations and how these differences can sometimes account for the human trait differences we observe. I will not opine on whether race exists or not; it does so in a very real sense. Instead, I will discuss what modern genetics says or does not say about the genetic meaning of race. Such work, together with increasing knowledge of how genes function at the molecular level, is giving us new insight into how genes influence our traits. But we remain vastly ignorant; so wild speculations about the genetic nature of many human attributes reside beyond the realm of today's science. Importantly, the notion of the gene, in the minds of most—many geneticists included—is one of an inflexible machine with deterministic outcomes. As the science advances, we increasingly find that the effects of genes are highly modifiable, dynamic, and subject to external influences (Chakravarti 2010). Indeed, why only genes? Even their ultimate products, structures such as our brain, are highly modifiable and dynamic (Kays et al. 2012).

---

[1] The word trait is used to indicate a distinguishing quality or characteristic; in its genetic flavor it means the "phenotype" or the manifesting feature of our genes ("genotype").

Cite this article as *Cold Spring Harb Perspect Biol* doi: 10.1101/cshperspect.a023358

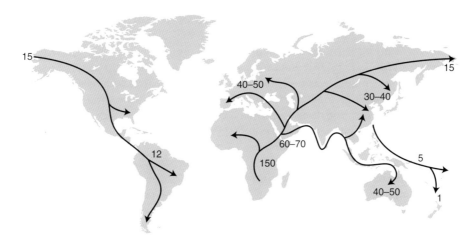

Figure 1. The long trek of our ancestors from the beginnings in Africa ~150,000 years ago to their emergence out of Africa about 50,000 years ago to colonize the Levant, Europe, Asia, Australia, Americas, and eventually, Oceania. (From Gluckman et al. 2009 [Fig. 6.6, p. 142]; reprinted, with express permission, from the authors in conjunction with Oxford University Press © 2009.)

Overlaid on top of all of this are the vast demographic changes human populations are increasingly experiencing, driven by increased communication and movement, and the shedding of past cultural divisions. These changes profoundly affect the distributions of genes across humanity and the melding of what were once population attributes. The science of genetics will be crucial to understanding how human history unfolds, and I predict that most of our current prejudices will turn out to have no biological basis.

WHY ARE WE NOT ALL THE SAME?

There would be nothing to argue about if humans were not different from one another. Of course, there are the rare exceptions of identical twins, but even though they have identical genomes they can on occasion show different traits. This perceptible difference within a collection of similar things extends across all of nature. Science is possible only because these differences exist and is driven by our continual quest to find out how and why they arise. Genetics is a young science, its 100-year history arising from the quest to find how and why biological differences arise and how they are maintained. The "how" was first answered by

Gregor Mendel's experiments with peas and the "why" by Charles Darwin's wondrous voyage to the Galapagos (Provine 1971). We persist in continuing to answer these questions in ever more detail because our current understanding, despite being solid, is very incomplete. We are acutely aware of what is not true but often on shaky ground about what is.

We differ in traits because of the biological processes that produce us. Some of this is genetically encoded and some of it is environmentally induced or modified. Each of us develops via a genetic program that is encoded by the sequence of A, C, T, and G bases in the DNA that makes up our genomes. Each of us inherits two genomes, one from each of our parents. This genetic program defines each of us uniquely and is more similar between any two members of the same species than between members of two related species. Thus, human genomes are more similar to one another than any one of our genomes to that of our great ape relative the chimpanzee. The differences do not stop there. Each of us is acutely aware of individual-to-individual differences between humans and the greater similarity between any one of us and our family members. Genetics provides a singular explanation for both observations. The closer the relationship between two persons the more similar their genomes and the greater the similarity of their traits. The dissimilarities between our genomes are owing to constantly arising mutations in our DNA. The majority of these are never transmitted to subsequent generations (they are lost), but a minority persists over time and across the generations. These surviving variations are the currency of modern genetics. Two individuals are related by virtue of sharing one or more common ancestors from whom they have inherited a small segment of their genome. The more remote the common ancestor(s), the less the fraction of the genome shared. Thus, we share 1/2 of our genome with each of our parents and siblings, 1/4 with each of our four grandparents, 1/8 with each of our first cousins, and so on (gcbias.org/2013/12/02/how-many-genomic-blocks-do-you-share-with-a-cousin).

The genetic mutations that persist among us, called genetic variants or polymorphisms, can occur at various frequencies within a population; some are rare and unique to a family, whereas others are common and have spread throughout humanity. These genetic variants can be assessed to evaluate how different two individuals are, and therefore figure out their relationship. The first human genetic variation identified was the ABO blood-group system in 1900, which was used almost immediately for assessing close relationships such as paternity. Today, technological advancements allow us to examine the entire genome in exquisite detail and to identify essentially all of such genetic differences. Typically, when one compares two copies of the human genome, say the maternal and paternal copies in any one of us, we find one of every 1000 bases to be different. Because the human genome is three billion bases long, that represents three million differences. Like the Hubble telescope that has allowed us to see deeper into space and time, new genomic technologies can identify all of these differences today and allow us to detect ever more remote relationships, and ancestries to 100,000 years before present or more, and do so on a global scale. The consequent stories of what these similarities and differences mean for similarities and differences in human traits and diseases are only in their infancy.

THE BIOLOGICAL BASIS FOR VARIATION IN HUMAN TRAITS AND DISEASES

Friar Gregor Mendel was the first geneticist (Orel 1996). He was deeply interested in the question of how differences in plant characteristics arise and how they are propagated. His now famous experiments, using simple observable traits (plant height, seed shape, flower color) of the pea plant, allowed him to infer that trait variants were owing to differences in separate "factors," that these factors existed as pairs in individuals, and that one of each was transmitted to each offspring, randomly and independently of other factors. Mendel's factors are today's genes. A fact not appreciated is that Mendel performed many other similar experiments with other traits and other plants and failed to uncover similarly clarifying principles (Orel 1996). This is not to say his rules of inheritance governing genes are incorrect; these apply to all genes. Rather, some genes have overwhelming effects on a trait and so the trait inheritance patterns are simple, the so-called Mendelian traits, such as those of plant height, seed shape, and flower color in the pea plant. Other genes, however, exert their effects in concert with numerous other genes, none of which overwhelm the others and thus have more complex patterns of inheritance. The fact is that Mendelian inheritance of "traits" is very much the exception, not the rule (Chakravarti 2010). The failure to understand this key feature led to considerable and bitter controversy in the early genetics literature (the Mendelian-biometrical debate) when some held that metrical traits such as height were not inherited but rather their variation arose solely from environmental differences (Provine 1971).

Eighty years hence, we are considerably more informed as to the nature of non-Mendelian inheritance. Much of the science of genetics has advanced from the experimental use of Mendelian inheritance to uncover its biological and molecular underpinnings. We also know that although some traits arise from the actions of two or at most a few genes, the vast majority of traits are genetically complex arising from the actions and interactions of numerous, hundreds or even thousands of genes. We know this because contemporary genetics allows us to map the locations of the individual genes contributing to a trait and, for most, a role for hundreds of genes has been revealed (Lango Allen et al. 2010). We also know that many more unmapped genes exist and that the individual-to-individual variation in a trait also involves the contributions from other domains, namely, environmental and epigenetic factors. The term environmental as used in genetics is both broad and nonspecific, and can include everything from lifestyle (diet, exercise) to ecological (weather, altitude) to social (income, education, healthcare access) and cultural (diet, belief systems) factors. The term epigenetic is also broad and includes a whole host of cellular processes that can direct the actions of genes without being dependent on the sequence of a specific genome. There is also increasing evidence that cellular (genetic) outcomes are not deterministic but inherently dynamic and stochastic. Genes do not determine only one outcome, but a range of possible ones. Finally, consider that biological effects cannot be arbitrary but are both canalized (restricted to certain possibilities) and built to preserve homeostasis (physiological regulation maintaining more or less constant internal conditions). There is one more arbiter; evolution decides which of the many genetic changes that occur within our genomes will be retained and which ones will be culled depending on whether the change is beneficial or not.

The variation in any trait, or for that matter disease susceptibility in any species including our own, is the result of many factors (genetic, epigenetic, environmental) each of which can be further divided into multiple subfactors. It is no surprise that traits can be inherited in a complex manner because beyond genes, whose inheritance patterns we understand, the epigenetic and environmental factors can also be "inherited," albeit according to rules still not understood (Cavalli-Sforza and Feldman 1981; Jirtle and Skinner 2007). It is then also unsurprising that Mendelian inheritance of traits is rare, because these represent the unusual singular effects of one gene that overwhelms the extant nongenetic variation.

Geneticists have long been interested in the precise genetic architecture of "complex traits." The first step is assessing a trait's genetic component. The top-down or classical approach has been to compare traits among relatives, because we have long known how much genetic information relatives share (e.g., 50% between siblings) without knowing individual genes. This allowed us to estimate the proportion of variation that is genetic, a proportion called heritability. The concept of heritability has been of great practical utility in plant and animal breeding, as a guide for choosing which strains to develop for improvement of yield and performance-related traits. The heritability of numerous human traits and diseases has also been measured, often repeatedly. Although some traits have high heritability, like height ($>80\%$), the overwhelming majority has low to moderate heritability (30%–50%) (Vinkhuyzen et al. 2013). The specific identification of the genes that explain this heritability, by contemporary bottom-up approaches in which the entire genome is systematically investigated, has been notoriously difficult, however (Lango Allen et al. 2010; Vinkhuyzen et al. 2013). This is sometimes referred to as the "missing heritability" problem.

There are many reasons for this apparent failure. First, our study samples are yet too small and not diverse enough. Second, our technological approaches are insufficient to recognize the vast network of gene interactions that may be principally important. Third, heritability estimates exaggerate the effect of genes because most studies cannot distinguish genetic from social or cultural sharing; family members share much more than genes (social, cultural, and dietary factors). Fourth, heritability is a relative measure of genetic versus nongenetic contributions. Thus, simply increasing or decreasing the environmental part of the variation can alter the apparent role of the genetic part. Well-known examples, such as the height increase seen from improved nutrition in the absence of any genetic change or the dietary treatment of phenylketonuria from birth to prevent intellectual disability, show that the actions of genes can be mitigated by nongenetic interventions.

WHY GENES TELL STORIES

Almost every human gene, when its genomic sequence is compared across individuals, shows variation. Each such sequence is a palimpsest, recording all changes from mutations that have survived until today. Some of these changes are unique to particular individuals, perhaps even one, whereas others are present in many of us. Because all humans belong to a single family tree—on average any two of us share 99.9% of our genomes—the fraction of sequence difference between any two genomes indicate how far back in time they had a common ancestor. Different segments of the genome are shared with different common ancestors; so the fraction shared or different between two genomes varies along its length. Consequently, examining the entire genome, as we can do today, is more informative than studying only one bit, such as the maternally inherited mitochondrial genome or the paternally inherited Y chromosome. The latter are informative nevertheless because they allow us to make inferences about our maternal and paternal lineages, respectively. Because all genetic changes accumulate over time, our genomes thus provide a history of how we, as individuals and as a species, came to be. Today we can compare individual genomes to infer our relationships, how far back in time we shared one or more common ancestors, and with increasing precision because of limited mobility of our ancestors, where our forebears were geographically located.

Genomic technologies, genomic sequencing in particular, have opened the door to recovering our individual and collective genetic histories, and, therefore, in concert with other sources of information, to uncovering details of our prehistory. In this sense, genes tell compelling stories about each of us as well as our shared humanity. This is a truly remarkable scientific and social achievement. There are many aspects of these stories that are uncertain and will require revision in the future. Nevertheless, some compelling and surprising truths have emerged. The most important of these is that contemporary humans are a remarkably young species and we all belong to a single family tree that arose from common ancestors a little more than 150,000 years ago (Pääbo 2014). Modern humans came to be in the last few minutes of the last hour of the last day if all 14.5 billion years of cosmic evolution were compressed into one year. If we were the common bacterium *Escherichia coli*, then this would correspond to a mere 3 months of our life. The story of human diversity, why we look as diverse as we seem to, needs to be told with this truth in mind.

WHAT IS RACE?

The Oxford English dictionary defines human races as the "major divisions of humankind, having distinct physical characteristics," and also as a "group of people sharing the same culture, history, language, etc." Biologists have had a more specific definition of race, one not conjured with human diversity in mind. The evolutionary biologist Ernst Mayr wrote that a race is "an aggregate of phenotypically similar populations of a species inhabiting a geographic subdivision of the range of that species and differing taxonomically from other populations of that species" (Mayr 2002). This definition has more to do with biogeography and taxonomy. However, there is an implicit assumption of both transmission and permanence of such taxonomy, and biologists impute the existence of some fundamental genetic and evolutionary difference between groups termed races. If one believes in evolution and modern genetics, and a common tree of life, the conclusion is inescapable that some members of a single species will be more different than others; additionally, close relatives of each of these members will be more similar to their closer rather than their more distant kin. It is unsurprising that this is true for humans and that our many attributes, including physical features, show this pattern. The precise pattern of sharing is a result of our specific evolutionary history and these differences are written in our genes and propagated through them. The construction and existence of human races in this regard, quite apart from social and cultural meanings, would not per se be controversial. It is controversial today because, over the past few centuries, both experts

Cite this article as *Cold Spring Harb Perspect Biol* doi: 10.1101/cshperspect.a023358

and nonexperts alike have brought in new and corrupted meaning that is not inherent in the biological concept. Discussions on human race, and caste, are difficult and incendiary today because their subtext is that human genetic differences are not neutral but either advantageous or disadvantageous and, tragically for human history, the corollary is that some groups have mostly advantageous attributes, whereas other groups have largely disadvantageous ones (Herrnstein and Murray 1994; Koenig et al. 2008; Wade 2014). Genetics has provided a second, more pernicious, corollary. Because some of these traits might be "genetic," these differences are transmitted at conception and so are biologically permanent (Herrnstein and Murray 1994; Wade 2014). The implication is that some groups have a genetic advantage. There have never been any empirical data to support these claims, and, moreover, the survival of "diverse" human groups is prima facie evidence of each of our groups' evolutionary success (Fraser 1995).

We humans must have always named and classified each other as we came across our evolutionary kin. However, the rise and expansion of the modern concept of human races had to wait for the conquests by colonial powers that brought Europeans into direct contact with many groups who were different with respect to language, culture, and even physical features. It should be remembered that this was always an asymmetrical and unequal rendezvous, favoring the colonizer and disfavoring the colonized. The notion of human races arose in this background from studies of comparative anatomy of human skulls by the German physician and naturalist Johann Blumenbach in the 18th and 19th centuries (Blumenbach 1775). He classified these skulls, and thereby humanity, into five major classes—Caucasian, Mongolian, Malayan, Ethiopian, and American—and began the horrid practice of providing color aliases (white, yellow, brown, black, and red). He correctly concluded that "individual Africans differ as much, or even more, from other individual Africans as Europeans differ from Europeans." Blumenbach, like his contemporaries, believed in the "degeneration hypothesis," which is that humans were originally "Caucasian" and that other races

were the outcomes of environmental degeneration (e.g., through exposure to sunlight). Despite the cultural biases he began with, Blumenbach was far more generous than his contemporaries with his view that Africans were not intellectually lesser than their European counterparts (Blumenbach 1775). Subsequently, despite other investigators classifying humans into anywhere from two to 63 races, no less an authority than Charles Darwin opined that "it is hardly possible to discover clear distinctive characters" between human races, because they "graduate into each other" (Darwin 1871).

The political and economic rise of Europe, and then the United States, in the 19th and 20th centuries, fed many bogus ideas into what evolved as "scientific racism" (Fredrickson 2002). There were parallel developments dealing with caste differences in India, although this is a much older classification. These studies had in common the examination of selected traits and attributes, deemed to be hereditary, in turn justifying the conclusion that there was a well-defined value hierarchy inherent in our species, with some groups much better endowed than others. The new emerging concepts of genetics added a new dimension; heritability assured that both the well- and less-well endowed continued to remain so. These beliefs—and they are largely beliefs of the perpetrators because so much of their data has been subsequently shown to be selectively used, manipulated, and outright fraudulent—led to a long period of eugenics both in the United States and Europe and the subsequent rise of the concept of inherent group superiority. This had disastrous consequences for Jews, Gypsies, intellectuals, nonconformists, and the mentally ill, among others, and biased immigration policies in many countries, including the United States. Unfortunately, geneticists had no small role to play in this crime of historical proportions (Witkowski and Inglis 2008). As a number of the authors in this collection describe, there is still a continuing tendency to conflate all manner of group differences with gene differences (Herrnstein and Murray 1994; Cooper 2013; Duster 2014; Wade 2014).

In the next section, I will turn to a modern accounting of human population variation and what it may say or not say about human races. Irrespective of that answer, one aspect is certain. Modern genetics research does not support the contention that one group or another has all of the positive traits, that we understand the genetics of complex traits sufficiently well enough to know that group differences in traits mean majorly group differences in genes, or that traits with a genetic component have fixed, inviolate, permanent, and unmodifiable effects. Even if we were to defend the idea of human continental ancestry or race it would be impossible to defend the assertion that they are inherently unequal. The remarkable feature of human evolution and adaptation is the widespread commonality of highly advantageous features (speech, cognition, culture) throughout humanity and the less frequent evolutionary innovations that occurred locally (pigmentation).

A BRIEF SYNTHESIS OF RECENT HUMAN EVOLUTION

The evolution of hominids leading to *Homo erectus*, 1.5 to 2.5 million years ago, and then *Homo sapiens* in Africa, is now well established. Although *H. erectus* existed outside Africa, the evidence is very clear that we are all descendants of groups from the African continent. *H. sapiens* first appeared there no earlier than ~300,000 years ago (Klein 1989). The subsequent history, evident in only fragmentary form through fossil remains, is where genetics has been indispensable (Cavalli-Sforza et al. 1994; Pääbo 2014).

The widespread discovery of gene variation in the 1960s immediately prompted studies to assess their relative relationships across human groups. The first study reconstructing human evolution using data from living groups was by Cavalli-Sforza and Edwards in 1964 (Edwards and Cavalli-Sforza 1964). This landmark study produced a "tree" in which extant populations arose through independent evolution by splitting from a common ancestral group that also produced a sister group, and so on. This study yielded two major findings beyond the specific relationships between groups. First, geographic proximity reflected greater genetic similarity across all groups, with the largest difference

being between African and Australian Aboriginal samples. This suggested that human colonization occurred through successive and serial migrations. Second, anthropometric measures and skin color showed a very different set of relationships, for example, a close association between African and Australian Aborigines, unrelated to geography but dependent on climate (Cavalli-Sforza and Edwards 1964). Another landmark study by Richard Lewontin in 1972 went on to show that the majority of human genetic variation, on average 85%, existed within any group and that intergroup differences were relatively minor, with the largest being between African and non-African groups (Lewontin 1972). These were not isolated controversial studies but rather the beginning of an onslaught of investigations, using successively larger and larger numbers of genes and humans, which have produced a single, consistent genetic narrative of human history (Cavalli-Sforza et al. 1994).

All early studies of human evolution compared the features and relationships of populations not individuals. In other words, these studies compared the relationships between frequencies of gene variants and not the genomes of individuals. This distinction is critical because past studies depended on the definition of a population. Is it defined by language, culture, geography, physical appearance, caste, or "race?" The definitions, of course, could skew the results one way or another. Of course, populations defined by known specific differences can be different at the genetic level. This is why Allan Wilson's 1987 study of individuals and their mitochondrial genomes is a significant departure from the past (Cann et al. 1987). His research accomplished four major goals. First, they studied individual genomes and not ensemble frequencies; second, they clearly showed that the human evolutionary tree had two major branches, one composed of African mitochondrial genomes only and the other comprising all humans including Africans; third, except for Africans, all other individuals from the same population had multiple origins; and fourth, they dated the common mitochondrial ancestor of all humans to be less than 200,000 years ago.

Cite this article as *Cold Spring Harb Perspect Biol* doi: 10.1101/cshperspect.a023358

There have since been many genetic studies using ever-increasing types and numbers of genetic variants and culminating in contemporary studies involving whole genome sequencing of diverse collections of humans. The essential conclusions of the findings described above have now stood the test of time. The single story of human evolution in the last 150,000 years is that all of us today are the descendants of early *H. sapiens* in Africa. A small group of these ancestors migrated "out of Africa" about 50,000 years ago and have since colonized the rest of our globe with each new group serially colonizing new unexplored geography. This is how we came into the Levant (Middle East) and then to both Europe and Asia and its subcontinent, and beyond into New Guinea and Australia. This is also how we went to remote parts of Siberia and came to colonize the New World about 15,000 years ago, which was until then hominid free. Our latest forays, only 2000 years ago, have been into Oceania (Fig. 1). These journeys provide the explanation for the pattern of quantitative differences that Cavalli-Sforza, Lewontin, and Wilson first brought attention to. Yes, there are differences in genetic variation at the continental level and one may refer to them as races. But why are continents the arbiter? If humans have had this single continuous journey disobeying continental residence—and as evidence we have the continuous distribution of genetic variation across the globe, not discrete boundaries like political borders—where do we divide humanity and why? (Weiss and Lambert 2014). All humans, without exception, are one species that has only very recently dispersed, with each population being more related to its proximal geographic neighbors. If we do look, behave, and have features that distinguish us markedly from one another, then these are differences that have arisen and amplified only over the last 50,000 years (2000 generations) and, quantitatively, are very small—only one part in 1000 bases. As a comparison, consider that chimpanzees and humans diverged from a common ancestor more than five million years ago (200,000 generations).

Of course, there are many more details to fill in and we wish to have far greater resolution of the history we already know. In this collection, scholars of genetic variation and evolution in the major geographic regions of the world have outlined both known and more recent genetic studies (Gomez et al. 2014; Majumder and Basu 2014; Ruiz Linares 2014; Veeramah and Novembre 2014). Their research, and the complementary works of others, allows us to make three major novel inferences. First, human populations were not always large and, probably, were almost always small. Second, the current abundance of a group is not a reflection of its past size. Third, human populations are seldom homogeneous and are highly admixed.[2]

We academics and nonacademics alike like to associate the genome only with its biological properties. However, our genomes and genetic variation between them and peoples are also the result of who lives, who dies, and who leaves behind how many offspring (i.e., demography). In other words, our genomes carry the record of both biology and demography, although teasing these apart is neither trivial nor easily corroborated by independent sources of information. In fact, contemporary whole genome genetic variation data emphasize the greater imprint of demography than biology in our genomes. As mentioned above, the chief conclusion of many genetic studies is that only a small group emerged from Africa to colonize the world and each successive colonization involved small numbers of founders as well (Gutenkunst et al. 2009), sometimes in the hundreds. This speaks to the limited amount of variation each founding group carried (which was also a subset of that in its parent group) and the constant threat of extinction to such a small band. Our eventual success is often chalked up to crucial adaptations; however, it is not implausible that there were many attempts, many catastrophes, and we are simply the lucky survivors. Of course, long-term survival came from population expansions, an intrinsic feature of human, and all other, evolutionary success. As other authors

---

[2]Admixture is used here in the genetic sense, in which extant genes and individuals arise from two different ancestries, as historically occurred with the colonization of the Americas by Europeans.

of this collection have documented, human population sizes rose within Africa at the time of our early ancestors, and then outside Africa as their descendants spread across the globe. These population increases must have depended on many environmental and chance factors as well as the inclusion of new arrivals (immigration). These factors are neither stereotypical nor orderly, once again emphasizing the strong effects of chance. A group populous today was not necessarily originally so; on the contrary, they may have been ever so close to extinction. Finally, although in the minds of many genetics is associated with homogeneity, human evolution is nothing else but a story of admixture and heterogeneity. There is now ample evidence in the genetic structures of the peoples of Africa, Asia, Europe, and the Americas that all extant humans are admixed (this collection). The new data are only now revealing whom we were admixed with, whether such admixture was common or rare, and when it occurred. Indeed, it is this story of mixing that is so at odds with the classical view of human group identity (Reich et al. 2009). We tend to think of admixture as a feature of modern times and, as has frequently happened, in terms of subjugation of one group by another as has occurred many times across human history and geography. However, this must have occurred even in the remote past; as human population density increased, there must have been a greater frequency of encounters with others and thus opportunities for mixing. The recent evidence that many humans carry genomic segments that can be traced back to Neanderthals and Denisovans (a second archaic human group) is evidence of such genetic exchange even 30,000 or more years ago with then-existing archaic human groups (Pääbo 2014).

The history that I have outlined above has implications for the natural selection and adaptation that have surely shaped our genomes. First, the major adaptive events that led to the emergence of *H. sapiens* are not the subject of debate. For matters of race, it does not even matter what happened to our young species in all of the last 150,000 years, or the emergence of modern humans while in Africa. What does matter are the adaptations in the past 50,000

years that have led to our spread across all continents and the genetic differences between us since then. This span of 2000 generations is long enough for specific adaptations to have occurred but not for very many such adaptations. The reasons, albeit technical, essentially depend on the following argument. Every adaptation occurs through some individuals possessing a beneficial mutation that, while bringing them an advantage (larger numbers of surviving offspring), is a relative disadvantage to those who do not carry that beneficial variant. This is a precarious gift because early in the evolution of this mutation most individuals are at a disadvantage and even those with an advantage might not realize their benefit. The English geneticist J.B.S. Haldane argued that this cost of selection was high enough that many beneficial mutations cannot arise at the same time (Haldane 1957). Of course, adaptations do occur as is evidenced by the striking examples of lactose persistence (selected after dairy farming arose) or skin pigmentation (selected for in response to solar radiation) (Quintana-Murci and Barreiro 2010). To me, even more striking is the repeated birth of the sickle cell mutation at the same DNA site in Africa as a protective response to malaria (Wainscoat et al. 1983). Nevertheless, these examples of adaptation are rare and there are likely only a handful of such examples in genomes across humanity (Hernandez et al. 2011). Positive selection and adaptations must have occurred, but their mechanisms likely are not through single genes but across many genes affecting the same trait, as is the case for height (Turchin et al. 2012). If that is the common scenario, then even strong selection on one gene among the many that affect a trait is going to be trivial unless we are speaking of very long evolutionary times or adaptive changes that occurred in our shared history.

The thesis that human groups substantially differ in most traits that are deeply rooted in simple genetics, and the result of recent adaptation, is fanciful. Human groups do differ from one another in many ways and the reasons are more likely to be nongenetic than genetic. Of those that are genetic, their composition is likely owing to many genes (hundreds to thousands),

Cite this article as *Cold Spring Harb Perspect Biol* doi: 10.1101/cshperspect.a023358

adaptations at many of them unlikely to be sustainable by known genetic mechanisms in the time frame during which human differences must have accumulated. More than 45 years ago, Motoo Kimura contended that, broadly, most of molecular evolution is deleterious and doomed to extinction; of those that do survive, the vast majority are selectively neutral (Kimura 1968). Recent data suggest this to be amply true. A benefit of this theory is that it means that the vast majority of changes in our genomes occur at a constant rate and provide an excellent "molecular clock" to date specific events in our common and unique histories.

THE NEXT PHASE OF HUMAN GENETIC DIVERSITY STUDIES

Human genetic diversity is dynamic and its patterns have changed substantially over time and will change in the future. Each of us can trace our genomes back to Africa and the subsequent journey across 50,000 years with intermixing with many other peoples. We live, however, in very different times. There are increasing rates of meeting and mixing, including groups that may have been relatively isolated for a few thousand generations. There are also rapid cultural and social changes that make neither gene nor cultural isolation possible. All of this implies even greater admixture than ever before. Consequently, I suspect the study of individuals and their genomes will increase at the expense of studying populations. It is remarkable how many individuals choose to study their genomes purely to decipher their ancestry (www.23andme.com). These individual genomes will surely uncover their individual histories but also begin to add detail to our common genetic history (1000 Genomes Consortium 2012).

I suspect that the focus on race, caste, or tribe that we still see today will erode simply because fewer and fewer members of any group will have its hallmark features. What does continental ancestry mean when one is from more than one? It might survive, I suppose; after all, how many fans of Manchester United across the globe have ever even been to Manchester? For a while, our ability to decipher individual histo-

ries might be useful to test how strict or porous the concept of a "population" is. Genetics and evolutionary biology have held as fundamental the concept that a population is a real, stable genetic unit, a property that is discrete and survivable. The reality is that most populations are dynamic and fluid, neither real nor stable.

Human evolution has always been studied with respect to such populations defined by language, geography, or cultural and physical features. Consider instead what we could decipher if we could sample a million humans (say), without regard to who they were, across a virtual grid across the world; this would correspond to sampling one person every 57 square miles ($\sim$7.5 miles $\times$ 7.5 miles) across the land surface. (This grid sampling idea was mentioned to me by the late Allan Wilson sometime around 1988.) Assume as well that we would sequence their maternal and paternal genomes and ask them questions, such as where they were born, where their parents were born, which language(s) they spoke, and what group affiliation(s) they had. We could then specifically uncover not only all of the features of human evolution we know today, and revise them to greater accuracy, but also test whether any or all of the features we use for human classification are supported by their genes. These types of global surveys of diversity have been performed for other species and may provide the first objective description of ours, bereft of race and other labels. This does not vitiate any social or cultural ways of defining humans, but at least one can no longer claim a genetic basis for all group differences.

RACE-BASED AND INDIVIDUALIZED MEDICINE

One of the major contributions of genetics to medicine has been, beyond the identification of disease pathophysiology, the recognition that each disease is multicausal and that patients with a single disease label may have conditions that arise from distinct molecular causes. This is well recognized for single gene disorders such as muscular dystrophy or even broad categories like prelingual deafness. Distinct molecular etiologies may require distinct therapies and management methods simply because physicians are trying to ameliorate different pathologies. These ideas have led to the concept of individualized medicine, that is, the tailoring of care to each patient depending on their genetic makeup and individual health circumstances (Childs 1999). This conceptual basis for care is a fundamental change in medicine that has long relied on the idea of a typological patient. Medicine's goal is now individualized care for the common chronic human diseases, a far more challenging task because the genetic and molecular bases for most chronic diseases remain unknown. We are making great progress but we have a long journey ahead before we understand the genetic, epigenetic, and environmental contributions to these disorders and which of these three may be the best route to intervention. That, of course, should not prevent us from individualizing care as best as we can with the knowledge we already have.

One of the major areas for individualized medicine is cancer therapy in which molecular diagnosis of germline mutations has been prevalent for more than two decades, and the genetic profile of the tumor has directed aspects of treatment. Increasingly, genome sequencing is being used to profile tumors broadly to direct treatment (Vogelstein et al. 2013). Individualizing risks to specific cancer subtypes and tumor profiling are expected to become routine aspects of cancer treatment particularly because personalized immune-modulation therapy is also on the horizon (Pardoll 2012). It is interesting to reflect that so much progress has been made in cancer treatment with greater discussion of personalized medicine rather than through the lens of race, despite cancer epidemiology data that show differences by continental ancestry. Differences in cancer incidence and prevalence by ancestry or ethnicity or community are well known. In fact, significant differences by geography, in the United States down to the county level, are also well known, suggesting major environmental etiologies as well. These differences, which lie beyond genes, need to be addressed simultaneously. The tussle lies in which genes and environments will we emphasize when both are responsible?

There is, of course, a long history of the study of variation in the incidence and prevalence of any disease by race, continental ancestry, and ethnicity. There is no doubt that many disorders show persistent and consistent differences and lead to great health disparities around the world (Murray et al. 2013). Genes do contribute to some part of it, probably no more than half, based on heritability studies, but environmental, epigenetic, and chance effects contribute significantly as well. Thus, equating all differences to genes is neither correct nor wise. As the essays by Richard Cooper and Troy Duster eloquently argue, the nongenetic factors in human disease, equally if not more importantly, affect our genetic biology in fundamental ways, and in many cases merit direct interventions that can lead to reductions in disease prevalence. Treatment of elevated blood pressure to prevent hypertension and its associated damage to the heart and kidney is a cogent example. In the United States, African Americans have elevated rates of hypertension and its sequelae as compared with those with European ancestry. This consistent finding has led to the myth that Africans have a higher genetic predisposition to hypertension, a fact clearly refuted by studies of many African communities whose blood pressures are lower than those of many European communities (Cooper et al. 2005). Moreover, recent genetic investigations in very large samples clearly show that blood pressure susceptibility variants detected in European ancestry subjects are also susceptibility variants for African, Asian, and South Asian subjects (Ehret et al. 2011). How could it not be so? Blood pressure regulation is a crucial human physiological trait under homeostasis, probably modulated by hundreds of genes, genetic variation of which was probably chosen in the early days of human evolution. It is not surprising that all humans likely share such variants. This is not to argue that additional variants did not arise later, variants that are consequently expected to show geographic clustering, but they are expected to be fewer. In other words, a race-based approach to medicine is a poor proxy for the type of genetic understanding we need to allow advanced medical interventions, like those available for cancer. Genetics can be far more useful in identifying the molecular underpinnings of human disease and treatment differences by focusing on all of humanity and including our diversity, not ignoring it (see Lu et al. 2014). Population differences do exist, but genes are not the sole agent for these differences, and nongenetic factors may have greater potency (Kahn 2013). An important corollary is that we should not speak of nature and nurture generally but take their respective roles on a case-by-case basis.

The challenge for understanding complex traits is thus considerable. Defining the roles of specific genes in the common chronic disorders of our time will lead to a much improved understanding of how and why a disorder develops (pathophysiology) and thus lead to improved therapies. And this progress will critically depend, I believe, on parallel progress in our understanding of how environmental and epigenetic factors impact our biology. These are the other sides of the genetic coin and there is no intellectual solution to one without the other. One of my mentors, Ching Chun (CC) Li, once remarked, "We are geneticists, not hereditarians; of course, the environment is important." As a plant breeder in his native China he had seen both genetically selected crops wither in a drought and the fantastical bogus claims of improved yields by Trofim Lysenko. We have to get the genetics content right this time. It is a fundamental biological challenge and an exciting one. It is even more exciting to figure out how the outside (environment) affects the inside (the genetic program) and, in parallel, how and why we as a species are so much more diverse in our cultures than in our genes. Vive la différence!

ACKNOWLEDGMENTS

I am indebted to Richard Cooper, John Novembre, and Richard Sever for critical comments on this paper.

REFERENCES

*Reference is also in this collection.

1000 Genomes Project Consortium. 2012. An integrated map of genetic variation from 1,092 human genomes. *Nature* **491**: 55–65.

Blumenbach JF. 1775. *De generis humani varietate nativa [On the natural varieties of mankind]*. University of Göttingen, Germany.

Cann RL, Stoneking M, Wilson AC. 1987. Mitochondrial DNA and human evolution. *Nature* **325**: 31–36.

Cavalli-Sforza LL, Edwards AWF. 1964. Analysis of human evolution. *Proc 11th Int Cong Hum Genet* **2**: 923–933.

Cavalli-Sforza LL, Feldman MW. 1981. *Cultural transmission and evolution: A quantitative approach*. Princeton University Press, Princeton, NJ.

Cavalli-Sforza LL, Menozzi P, Piazza A. 1994. *The history and geography of human genes*. Princeton University Press, Princeton, NJ.

Chakravarti A. 2010. *Principia genetica*: Our future science. *Am J Hum Genet* **86**: 302–308.

Childs B. 1999. *Genetic medicine: A logic of disease*. Johns Hopkins University Press, Baltimore.

* Cooper RS. 2013. Race in biological and biomedical research. *Cold Spring Harb Perspect Med* **3**: a008573.

Cooper RS, Wolf-Maier K, Adeyemo A, Banegas JR, Forrester T, Giampaoli S, Joffres M, Kasterinen M, Primastesta P, Stegmayr B, et al. 2005. An international comparative study of blood pressure in populations of European vs. African descent. *BMC Med* **3**: 2.

Darwin C. 1871. *The descent of man and selection in relation to sex*. John Murray, London.

deGrasse Tyson N. 2014. *The Colbert Report*, March 10, 2014.

* Duster T. 2014. Social diversity in humans: Implications and hidden consequences for biological research. *Cold Spring Harb Perspect Biol* **6**: a008482.

Edwards AWF, Cavalli-Sforza LL. 1964. Reconstruction of evolutionary trees. In *Phenetic and phylogenetic classification* (ed. Heywood VE, McNeill J), pp. 67–76. The Systematics Association, London.

Ehret GB, Munroe PB, Rice KM, Bochud M, Johnson AD, Chasman DI, Smith AV, Tobin MD, Verwoert GC, Hwang S-J, et al. 2011. Common polymorphisms impacting blood pressure and cardiovascular disease in diverse populations highlight novel biological pathways. *Nature* **478**: 103–109.

Fraser S. 1995. *The Bell curve wars: Race, intelligence, and the future of America*. Basic Books, New York.

Fredrickson GM. 2002. *Racism: A short history*. Princeton University Press, Princeton, NJ.

Gluckman P, Beedle A, Hanson M. 2009. *Principles of evolutionary medicine*. Oxford University Press, Oxford.

* Gomez F, Hirbo J, Tishkoff SA. 2014. Genetic variation and adaptation in Africa: Implications for human evolution and disease. *Cold Spring Harb Perspect Biol* **6**: a008524.

Gutenkunst RN, Hernandez RD, Williamson SH, Bustamante CD. 2009. Inferring the joint demographic history of multiple populations from multidimensional SNP frequency data. *PLoS Genet* **5**: e1000695.

Haldane JBS. 1957. The cost of natural selection. *J Genet* **55**: 511–524.

Hernandez RD, Kelley JL, Elyashiv E, Melton SC, Auton A, McVean G, 1000 Genomes Project, Sella G, Przeworski M. 2011. Classic selective sweeps were rare in recent human evolution. *Science* **331**: 920–924.

Herrnstein RJ, Murray C. 1994. *The Bell curve: Intelligence and class structure in American life*. Free Press, New York.

Jirtle RL, Skinner MK. 2007. Environmental epigenomics and disease susceptibility. *Nat Rev Genet* **8:** 253–262.

Kahn J. 2013. *Race in a bottle*. Columbia University Press, New York.

Kays JL, Hurley RA, Taber KH. 2012. The dynamic brain: Neuroplasticity and mental health. *J Neuropsych Clin Neurosci* **24:** 118–124.

Kimura M. 1968. *The neutral theory of molecular evolution*. Cambridge University Press, Cambridge.

Klein RG. 1989. *The human career: Human biological and cultural origins*. University of Chicago Press, Chicago.

Koenig BA, Lee SS-J, Richardson SS. 2008. *Revisiting race in a genomic age*. Rutgers University Press, Piscataway, NJ.

Lango Allen H, Estrada K, Lettre G, Berndt SI, Weedon MN, Rivadeneira F, Willer CJ, Jackson AU, Vedantam S, Raychaudhuri S, et al. 2010. Hundreds of variants clustered in genomic loci and biological pathways affect human height. *Nature* **467:** 832–838.

Lewontin R. 1972. The apportionment of human diversity. *Evol Biol* **6:** 391–398.

* Lu YF, Goldstein DB, Angrist M, Cavalleri G. 2014. Personalized medicine and human genetic diversity. *Cold Spring Harb Perspect Med* doi: 10.1101/cshperspect.a008581.

* Majumder PP, Basu A. 2014. A genomic view of the peopling and population structure of India. *Cold Spring Harb Perspect Biol* doi: 10.1101/cshperspect.a008540.

Mayr E. 2002. The biology of race and the concept of equality. *Daedalus* **131:** 89–94.

Murray CJ, Vos T, Lozano R, Naghavi M, Flaxman AD, Michaud C, Ezzati M, Shibuya K, Salomon JA, Abdalla S, et al. 2013. Disability-adjusted life years (DALYs) for 291 diseases and injuries in 21 regions, 1990–2010: A systematic analysis for the Global Burden of Disease Study 2010. *Lancet* **380:** 2197–2223.

Orel V. 1996. *Gregor Mendel: The first geneticist*. Oxford University Press, Oxford.

Pääbo S. 2014. The human condition—A molecular approach. *Cell* **157:** 216–226.

Pardoll DM. 2012. The blockade of immune checkpoints in cancer immunotherapy. *Nat Rev Cancer* **12:** 252–264.

Provine WB. 1971. *The origins of theoretical population genetics*. University of Chicago Press, Chicago.

Quintana-Murci L, Barreiro LB. 2010. The role played by natural selection on Mendelian traits in humans. *Ann NY Acad Sci* **1214:** 1–17.

Reich D, Thangaraj K, Patterson N, Price AL, Singh L. 2009. Reconstructing Indian population history. *Nature* **461:** 489–494.

* Ruiz Linares A. 2014. How genes have illuminated the history of early Americans and Latino Americans. *Cold Spring Harb Perspect Biol* doi: 10.1101/cshperspect.a008557.

Smithsonian's National Museum of Natural History. 2013. Panel discussion on "ancestry and health." National Museum of African American History and Culture, National Human Genome Research Institute, September 12, 2013.

* Thapar R. 2014. Can genetics help us understand Indian social history? *Cold Spring Harb Perspect Biol* doi: 10.1101/cshperspect.a008599.

Turchin MC, Chiang CW, Palmer CD, Sankararaman S, Reich D, Genetic Investigation of ANthropometric Traits (GIANT) Consortium, Hirschhorn JN. 2012. Evidence of widespread selection on standing variation in Europe at height-associated SNPs. *Nat Genet* **44:** 1015–1019.

* Veeramah KR, Novembre J. 2014. Demographic events and evolutionary forces shaping European genetic diversity. *Cold Spring Harb Perspect Biol* doi: 10.1101/cshperspect.a008516.

Vinkhuyzen AA, Wray NR, Yang J, Goddard ME, Visscher PM. 2013. Estimation and partition of heritability in human populations using whole-genome analysis methods. *Annu Rev Genet* **47:** 75–95.

Vogelstein B, Papadopoulos N, Velculescu VE, Zhou S, Diaz LA Jr, Kinzler KW. 2013. Cancer genome landscapes. *Science* **339:** 1546–1558.

Wade N. 2014. *A troublesome inheritance: Genes, race and human history*. Penguin, New York.

Wainscoat JS, Bell JI, Thein SL, Higgs DR, Sarjeant GR, Peto TE, Weatherall DJ. 1983. Multiple origins of the sickle mutation: Evidence from βS globin gene cluster polymorphisms. *Mol Biol Med* **1:** 191–197.

* Weiss KM, Lambert BW. 2014. What type of person are you? Old-fashioned thinking even in modern science. *Cold Spring Harb Perspect Biol* **6:** a021238.

Witkowski J, Inglis J (eds.). 2008. *Davenport's dream: 21st century reflections on heredity and eugenics*. Cold Spring Harbor Laboratroy Press, Cold Spring Harbor, NY.

What Type of Person Are You? Old-Fashioned Thinking Even in Modern Science

Kenneth M. Weiss and Brian W. Lambert

Department of Anthropology, Penn State University, University Park, Pennsylvania 16802

Correspondence: kenweiss@psu.edu

People around the world have folk origin myths, stories that explain where they came from and account for their place in the world and their differences from other peoples. As scientists, however, we claim to be seeking literal historical truth. In Western culture, typological ideas about human variation are at least as ancient as written discussion of the subject, and have dominated both social and scientific thinking about race. From Herodotus to the Biblical lost tribes of Israel, and surprisingly even to today, it has been common to view our species as composed of distinct, or even discrete groups, types, or "races," with other individuals admixed from among those groups. Such rhetoric goes so much against the well-known evolutionary realities that it must reflect something deep about human thought, at least in Western culture. Typological approaches can be convenient for some pragmatic aspects of scientific analysis, but they can be seductively deceiving. We know how to think differently and should do so, given the historical abuses that have occurred as a result of typological thinking that seem always to lurk in the human heart.

Every naturalist who has had the misfortune to undertake the description of a group of highly varying organisms, has encountered cases (I speak after experience) precisely like that of man; and if of a cautious disposition, he will end by uniting all the forms which graduate into each other, under a single species; for he will say to himself that he has no right to give names to objects which he cannot define.

—Charles Darwin
Descent of Man, 1871

Do races exist? We often hear scientists, pundits, and even just ordinary people debate this question, usually contentiously, and collections (such as this one) are published to try to answer it. But the answer is clear and very simple, and it is: yes, races exist!

There can be no doubt about this. Too many people use the word routinely, and this by itself gives the term, and their lives, some existential meaning, and by default some sort of empirical meaning as well because it can affect where and how they live. Yet the variation in the word's usage makes it equally clear that the concept of "race" exists separately in the mind of each beholder. The legitimate scientific issue is the meaning and utility of the term in human biology.

Why do we still need repeated iterations of angst-laden discussions about what, or even if, "race" is? The relevant facts were known at the origins of modern scientific treatments of this subject, with caveats and clear statements made repeatedly ever since (e.g., see Kittles and Weiss

2003; Weiss and Fullerton 2006), including Darwin's own famous statement, given in our introductory quote. The persistence of the issue suggests that there are no experts who can enlighten us, with any finality, which suggests that this is not a matter of epistemological expertise. The reasons have little to do with the underlying biology: The reasons are cultural. However, they also entail concepts of human history and that is a subject to which one might expect that genetics could contribute in some definitive way. Indeed, the history of those concepts of history themselves is informative and worth a brief summary.

A BIT OF HISTORY

Not so long ago, it was not at all clear that the living world had much of a history. Species, as scientists, scholars, and the ordinary person knew them, had been around, essentially unchanged, since the earliest recorded writings. A set of instantly created species was the Biblical explanation (in the West, at least), because, after all, how else could oats, goats, whales, and snails have gotten here? Although here and there one can find alternative speculations, the panoply of Nature's species seemed, from the ancients to the 19th century, to constitute a complete fabric of the possible and useful existence, with a hierarchy of types of animals and plants, from simple to complex with humans at the top, the pinnacle of existence. These types were widely seen as permanent and essentially unchanging categories of nature, God's wisely constructed spectrum that came to be called the Great Chain of Being (Lovejoy 1936). In this context, in the middle 1700s, Carl Linnaeus developed a system of classifying living organisms that we still use today.

The major transformation in thinking was the development of a scientific rather than religious explanation for the origin of species, which Darwin referred to as the "mystery of mysteries." The genius of Darwin and Alfred Wallace (and a few precursors) was to see that the same variety of types of beings could be generated by historical "processes" alone, without the need for discrete "events" of spontaneous generation or divine creation. The process

was the gradual one that we call "evolution." In essence, types led to other types by gradual changes over eons of time, and all types that are here today resulted from that process operating since, and everything descended from a single origin of life on earth. That challenged religious comforts, but provided the ability to understand the world in scientific terms.

The concept of natural types is closely connected historically to the notion of a "type specimen," assumed to be representative of their kind, and museums were staffed (and stuffed) with them. Even well after Darwin, however, there was discussion of what constituted a type and how many individuals, or which individuals, were needed to define one (e.g., Schuchert 1897). Type specimens are distinct discrete entities, taken as representative because everyone has known that variation is the essential key to evolution. In his studies of barnacles, an 8-year drudgery that he undertook in part to build support in his own mind for his theory of evolution, Darwin observed that every part of every species varied. He bemoaned that "Systematic work would be easy were it not for this confounded variation" (Darwin 1850). That must be the case if adaptive evolution were to occur, because evolution requires variation, which is in turn necessary for populations to split into other populations that over time become reproductively isolated—and become new species. This, in its turn, challenged the idea of the static fixity of species on which pre-evolutionary ideas rested and that motivated Darwin's innovative thinking. And there is no hierarchy in this process; bacteria are still here and doing very well after 3.5 billion years! Darwin showed that the appearance of species stasis was an illusion due to the glacially slow changes that evolution wrought.

OUR PLACE IN NATURE

In subtle ways, even scientists, who know better, as well as the general public, still indulge in careless typological thinking about humans as well as other species. Western culture has conflated concepts of type and race for obvious reasons that we might call historical. Races as

 Cite this article as *Cold Spring Harb Perspect Biol* doi: 10.1101/cshperspect.a021238

natural types were inferred in pre-Darwinian times as God's separate creations, and although they were interfertile they were considered equivalent to subspecies. Even well into the 20th century, descriptive rhetoric often has taken such forms as "the Caucasian has a broader head than the Negro . . ." or a more euphemistic version "Europeans have . . . ." This is not really different from how we refer to the mouse, horse, or snapdragon.

Darwin's quote shows that it was well recognized even then that whatever these human "types" were, they were not really fixed types. Yet, great effort has been made to identify the human types by generations of anthropologists who professed to be evolutionists, even after stating the brief caveat that variation is quantitative and overlapping, the caveat immediately thereafter honored in the typological breach.

The most important point is that if one is determined to identify human types in the real, rather than Platonic ideal sense, one must assume that they actually exist to obtain specimens of them. In a kind of circular logic, the belief in the existence of races places a statistical bias or prior probability on the number or identity of races, assumed to be real ontological entities, which samples collected on that basis are then used to reinforce. We do that by prejudging the question and, for example, by sampling from discrete locales (populations, languages, ethnicities, etc.). In other words, to assess the variation between populations, we have to identify them, often by name ("Europeans," "Asians," "Nigerians"), which itself then determines the samples that are collected. One can always find statistical differences between any kinds of samples, and there are certainly practical issues involved in sample choice for anthropology and genetics, but sampling from preidentified populations almost enforces categorical interpretation, as if the sample choice were dictated by the categories. But sample choice is subjective, and if we force it into our assumed conceptions, we can make the statistical shoe fit. This is especially problematic when the objective extends beyond categorical treatment of present populations to using such concepts to reconstruct individual genetic ancestry.

ANCESTRY TESTING AS TYPOLOGICAL THINKING

Genetic ancestry testing has become a boom business. It is marketed as popular recreation, supported by appealing television documentaries, and widely used in science as well. It appears to be positively viewed by those who used, or who would use, these services (Wagner and Weiss 2012). Besides the fun of getting an idea of who one's ancestors were, ancestry estimation can have epidemiological relevance. Genetic ancestry testing uses an individual's genotype (variant nucleotides in his/her DNA sequence) to estimate the fractions of that genotype that was derived historically from a set of putative "ancestral" or "parental" populations, usually referring to geographic regions, whose frequencies of a set of tested alleles (genetic variants) are known. Often, although the word itself may not be used, these regions correspond to the homelands of the classically denoted human races. The test individual is said to be the "admixed" descendant from those populations.

Besides ancestry estimation of individuals requesting the service, admixture-based concepts are also routinely used to describe the ancestral history of samples of multiple individuals from present-day populations, such as U.S. African-Americans (Shriver and Kittles 2004), Hispanics and Native Americans (Wang et al. 2007, 2008; Bryc et al. 2010), Africans (Tishkoff et al. 2009), Europeans (Bauchet et al. 2007; Novembre et al. 2008), and South Asians (Reich et al. 2009). Indeed, variation in our entire species has been analyzed in this way (Rosenberg et al. 2002; Li et al. 2008).

Admixture approaches to human biological diversity take as an assumption the reality of the parental populations; that is, it is assumed that there are, or were, such "pure" human populations, and that everyone is either a member of such a group, or is admixed from them. Historical parentals are assumed to be represented accurately by sampling some current population.

In this way, whether or not the term "race" itself is used, prominent and sophisticated analysis of population history continues to be based fundamentally on racial (although not racist) assumptions. But if parental populations actually exist (or existed), we must identify them somehow, decide where and how many they are (or were), and explain how individuals can be assigned ancestry from them. That turns out not to be so easy.

STATISTICAL DEFINITIONS OF POPULATION AND TYPE

We should clarify what we mean by a population, whether it is admixed or not. Otherwise, the word can carelessly be used as a euphemism for "type," which can mislead one into thinking that a population is a collection of individuals all of the same "type" and isolated from other such collections. Instead, in the admixture estimation context, population is a different and rather subtle kind of type.

A typical definition of such a population is a collection of individuals who choose their mates randomly from among others within the population, but not from other populations. Extending this idea, an admixed population is one that was formed by contributions from two or more such "parental" (donor) populations, but that at the time of observation has random mating internally.

The idea of random mate choice may seem strange to readers not familiar with the genetics of populations, but one should not worry about that because the concept is an effective working simplification, which essentially just means that mates are chosen from within the population without regard to the choice's particular genotype. That this is not literally true can cause some problems, as we will see, but for the moment it does not affect the points we wish to make.

The view of humans as being members of self-contained populations is, however, rather strange because neither the facts nor theory provide support for the idea of rigid population boundaries in the process by which actual human variation has been generated and distributed around the Earth. Traits such as language and religion have always posed some local restrictions on mating, but by no means were they complete nor did they lead to truly closed, random-making populations. In this sense, the common approaches that invoke discrete ancestral populations, although couched in evolutionary terms, are basically non-Darwinian (Weiss and Long 2009). Decades of anthropology have shown that populations exchange mates often through mandatory village exogamy in which mates must be chosen from members of some group other than one's own. Even geographic barriers are typically not complete (except perhaps some truly isolated and relatively recent habitation of mountain valleys, Pacific islands, etc.).

Admixture-based analysis uses a particular kind of typology that recognizes that members of a "pure" population, like A and B in Figure 1, are not clones; their individuals vary. What is actually suggested, if only implicitly, is a "statistical typology," in which the population is in random-mating proportions for the frequencies of alleles in the population, at a set of sites

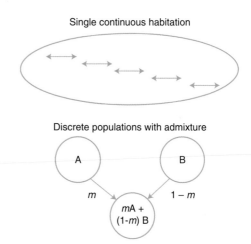

Figure 1. Populations and admixed populations. (*Top*) The human population as a quasi-continuous range based on serial expansion from a single source. (*Bottom*) A population formed by a fraction, m, of immigrants from discrete population A and the remainder from population B (for description, see text).

Cite this article as *Cold Spring Harb Perspect Biol* doi: 10.1101/cshperspect.a021238

to be examined. If different members of a population have different nucleotides (A, C, G, or T) at a given position, an individual's genotype at that site is the pair of alleles that s/he carries at that site. In a random-mating population, each individual draws its genotypes from the same set of possibilities—the allele frequencies for each of the varying sites to be considered. For example, suppose we consider a site with two nucleotides present in the population, say, 20% A and 80% G. If there is random mating, the chance someone is AA is 16% (20% for the person's first copy and, again, 20% for its second). A similar story would apply to every other varying site, applying that site's allele frequencies to the probability that the individual had each of its possible genotypes, and so on across the genome. For readers familiar with the term, the population is in Hardy-Weinberg equilibrium for the respective allele frequencies.

What constitutes a population in the current sense is that the variation among its members is in these random-mating proportions. These are the statistical "types" that comprise the population; the "type" is, in a sense, the set of allele frequencies from which each member's unique genotype was produced.

Individuals in an admixed population are treated as having drawn their genotypes as random samples of alleles from the respective allele-frequency sets of their contributing parental populations, weighted by the proportion of admixture (e.g., m in Fig. 1). In the admixed population, the genotypes are also due to random mating, but with these admixture-weighted allele frequencies. The effect of admixture is similar to mixing paint of different colors. The relative amounts of red and white paint that formed some new mixed paint would determine its shade of pink.

Something about these ideas might seem strange, and it is important to be aware of them. Genotyping in an admixture-based analysis is typically restricted to globally varying sites, that is, sites in which the same alleles are found in many or even all of the parental populations, although their allele frequencies may vary among the populations. This means that by the very assumptions of admixture analysis it is possible for people in any of the parental populations to have precisely the same genotype, yet those populations are treated as different "types!"

It may seem curious to define distinct parental populations in terms of alleles they all share, but it is pragmatically important. Each newborn person, no matter where, inherits new mutational variants that are not found anywhere else. It is not very helpful to use such variants in comparing groups, and indeed one else in the same group has a new variant. Similar uselessness applies to variants that are quite rare in any group. The poet John Donne said that no man is an island, but if one only used each man's unique alleles it would be much more difficult to identify groups; the concept of variant frequency would lose much of its meaning, as would the genetic concept of population itself. We are, thus, constrained to compare group differences by things whose variation is shared among the groups. And, once we have a sample divided into such populations, or sets of statistical types, despite the fact that in principle any genotype could be found in any population, if we look at enough sites we can always assign an individual to his/her respective population. This is a curious result of combining many different probabilities (from the array of tested sites in the genome). Although it is *possible* for any given genotype to arise in any population, if enough sites are considered the *probability* of that person's genotype arising in any population other than his/her own becomes miniscule.

This kind of admixture-based analysis was initially developed at least in part not for direct investigations of true population history in the ancestry sense, but to detect structure within samples used for gene-mapping studies to reduce false-positive associations between single-nucleotide polymorphisms (SNPs) and diseases. Substructure within a population can lead to such results, suggesting that some place in the genome contributes to the disease, which can mislead follow-up clinical research. For example, a subgroup might share a disease and also (by chance) some genetic variant that has nothing to do with the disease, but the association between the two could look like causation if the existence of the subgroup was not taken into account.

For this reason, the admixture approach is now often called a "structure" analysis after the name of the first modern computer analysis program that was based on this approach (Pritchard et al. 2000), of which there are now others (e.g., Tang et al. 2005a; admixmap.sourceforge.net). Because even within local villages, humans do not literally choose mates at random, random mating is a pragmatic statistical concept rather than one that attempts to address actual history, and more fine-grained analysis shows that populations treated as homogeneous at one level of resolution have comparable internal diversity at more local levels of resolution (Novembre et al. 2008; Wang et al. 2008; Weiss 2010). But the fact that a population's internal admixture structure depends on how closely you choose to examine it, casts questions about the historical validity and applicability of the approach.

It is important to note that categorical treatment of human variation, including implicit statistical race definitions, is not new (Kittles and Weiss 2003; Weiss and Fullerton 2006; Weiss and Long 2009; Weiss 2010). Because of their sorry historical abuse, words like "pure" and "race" are not commonly used by scientists today. But euphemistic terms like "ethnic group" are, and if we are aware of the problems then why are we doing the same kind of conceptual analysis after a 150-year history of evolutionary reasons to know better?

FITTING A SQUARE CONCEPT INTO A ROUND REALITY

How does a categorical view square with the observed ubiquitous, more or less continuous, human geographic variation? By the 20th century, more modern concepts than classical Linnaean static types were available and attempts were made to fit a discrete typology into a more continuous whole. For example, the leading physical anthropologist, E.A. Hooton, published a sober discussion in *Science* in 1926 on methods for analyzing races (Hooton 1926) that exemplifies this type, so to speak, of thinking. Hooton said

that to characterize human races in modern scientific terms, one must metrically analyze collections of (say) skeletons, first grouping them into clear sets (by using expert judgment!). Then, he argued, with the multivariate techniques of the time, the patterns of variation within and between groups could be discerned quantitatively; pure races will be "very apparent." These are the product of "evolutionary factors," he wrote, and with this as a basis one can use metric techniques to identify the "composite" nature of other groups produced by intermixture among these primary races. Here, we have what seems on the surface to be a modern, Darwinian, process-based way to consider types.

Hooton was a physical anthropologist, but he was writing at the dawn of genetics, and geneticists jumped into the fray to modernize this subject (or, as they would proclaim, to make it more rigorous). Because genes are taken as the fundamental causal units of life and evolution, it seemed to some that races should be defined by sets of clearly Mendelian traits (that is, that are each inherited as if because of variation in a single gene) rather than much "softer" morphometric data that may be affected by environments (Mendelian segregation of phenotypes was the criterion for genetic causation at that time when specific underlying genes were not known). This was the approach taken by the globally leading textbook in human genetics (Baur et al. 1931).

Baur et al. (1931) argued that races defined in this way are "sharply delimited" and "there are only men and women belonging to particular races or particular racial crossings." It is more than incidental that this book was squarely within the eugenics era, and viewed racial traits as the product of Darwinian selection. The Darwinian perspective initially led to discrimination against individuals deemed by physicians to be inherently deficient. But in a typological age, it was easy to extend the same ideas to value judgments about inherent characteristics of groups, and such ideas provided a feeder justification for the Nazis. The history of eugenic abuses is beyond our scope here, but the issues have been reviewed elsewhere (Kevles 1995; Carlson 2001; Weiss and Lambert 2010;

Cite this article as *Cold Spring Harb Perspect Biol* doi: 10.1101/cshperspect.a021238

and see articles in the May 2011 *Ann Hum Genet*, Vol. 75, No. 3).

All the eugenic-era investigators were well, indeed explicitly, aware of variation within their "types" to the point that no two non-twin members of the same type were identical. A subtle innuendo was that this variation did not overlap between groups in any substantial way, but this clearly is not true if the alleles are, by choice, present in different groups. Clearly as well, results were then (and remain today) dependent on the samples chosen for study. If one only samples Europeans and Asians to define one's races, their chosen traits might not overlap with, say, Indians, who could then be viewed as admixed between Europeans and Asians. But if one were to choose Africans and Europeans as parent populations, Asians could be viewed as their admixed descendants.

Nonetheless, and if we can blinker ourselves to overlook the abuses that took place in the name of such thinking, and its misappropriation of Darwin's name, we can see the sleight of hand that is involved. What we have is a definition of a round peg in a square hole: a variable type.

As described above, in population genetic terms, a variable type (a statistical "race") is de-fined as a population whose individuals were formed by randomly drawing variants twice from each test location in the genome. And also as described above, this is what the admixture-based approaches do as well, given the estimated mixing population proportions. In the admixed case, this means that each individual can by chance have drawn somewhat more, or fewer, variants that had come originally from a given source population.

Figure 2 shows a common graphical portrayal of an admixture-analysis result, in which sampled individuals are arrayed along a linear axis, grouped according to the sample from which each individual was obtained; the groups usually arrayed in geographical order, such as from west to east. The analytic software identifies (or is asked to identify) a number of ancestral ("parental") populations, each of which is given a color code. Then every sampled individual is plotted as a thin vertical bar, with color-coded segments corresponding to the parental populations and of length proportional to the estimated fraction of the individual's ancestry from that parental population.

In the resulting figures, geographic proximity is clearly reflected by the similarities of admixture patterns in individuals sampled from

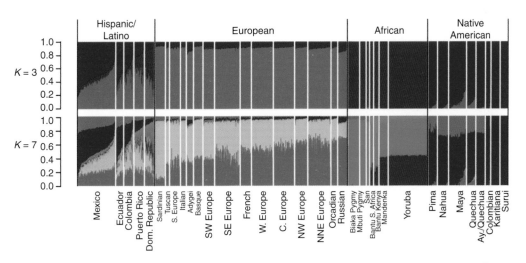

Figure 2. An admixture-structure presentation of a global sample of human genetic variation. Ancestry fraction scale is on the left. The *top* bar assumes three ancestral populations ($K = 3$), of which all individuals are members or admixed descendants, whereas the *bottom* panel is the same if seven ($K = 7$) ancestrals are assumed. (From Bryc et al. 2010; in the public domain.)

the same or nearby geographic regions (these individuals are thus adjacent or near to each other along the plot). If the number, K, of random-mating parental populations is to be estimated by the program rather than prespecified by the investigator, the number of parental populations becomes a matter of judging the statistical results, and the investigators choose what seems to be the best number (papers usually present results from various tested K-values, specifying which they feel is best). The typical global- or continent-scale study presents values between $K = 5$ and 15. For some individuals, the program assigns virtually all of their ancestry to a single parental population, essentially meaning that their genotypes are "pure" representatives of that population. But most individuals are estimated as having ancestry from two, or even more parental population sources.

This is clearly a literal fiction because the sampled individuals are all contemporary, may live very far apart, and each person can have only two immediate parents. In this sense, the admixture methods confound direct genealogical with population-historic concepts because the evidence must reflect earlier generations of contribution. There is no surprise in that, but it does imply that the contemporary sample interpretably represents assumed ancestral parental populations that really did exist as such. This assumption entails vague mixing concepts, imposing boundaries based on current data, and/or on populations assumed to have existed as discrete evolutionary units at some unspecified point in the past. Because the sampled individuals live geographically very far apart, the gene flow that is assumed must have had at least some historical depth, implicitly extending the assumptions about the long-standing purity of the parentals. The depth of history being reflected in this kind of analysis is rarely stated, indeed, would be very problematic to state convincingly, because it depends on the samples chosen and how they are interpreted. African Americans, for example, have some African and some European (and possibly other) ancestry, but that could have come from various places in the different continents and one or some unknown number of times in the 500 years since Columbus "discovered" the Americas (or by pre-Columbian European/African contacts).

Because to do this kind of analysis one must define the populations and the sampling frame, the analysis is often dangerously close to circular or self-affirming. It is, of course, easy to identify groups for sampling by language, political unit, nation, or because some anthropologist decided to live there for a while and gave them a name. Thus, in Figure 2, if there are assumed to be only three global contributing populations, the admixture pattern in each individual becomes simpler than if one assumes seven parental populations, but of course there has been only one actual human population history.

How these criteria reflect real history as well as the analysis itself may vary with the software program used in the admixture analysis, and each program has its own assumptions and methods, which can affect the results. For example, the investigators of the original modern admixture program called STRUCTURE (Pritchard et al. 2000) clearly provide all of the appropriate caveats, in clear terms, and most importantly the subjective judgments required in interpretation and that the program is designed to estimate admixture history even when admixture is an appropriate way to view what actually happened (if that is even known).

The presence of such cautions does not imply they are heeded or clearly acknowledged in the papers reporting use of the programs. Admixture analysis makes nice stories, although we know they are fairy tales, and the availability of convenient statistical programs does not justify users of such programs to present results that are manifestly misleading. For example, if one restricts one's geographic attention and looks close-up within a putative parental population by subsampling in its home region, one finds similarly rich internal admixture structure (structure in the same sense and revealed by the same kind of analysis) as was found in the larger geographic area. Figure 3 shows this clearly with regard to Europe relative to Europeans considered in the global context in Figure 2; such intraregional heterogeneity is, of course, widely recognized even by the proverbial man

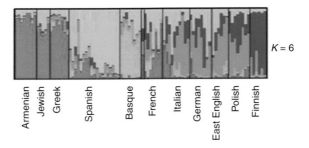

Figure 3. A geographic close up of one region, Europe, covered in Figure 1. (From Bauchet et al. 2007; reprinted, with permission, © Elsevier.)

in the street. Note that in Figure 2 we presented results comparing whether $K = 3$ or 7 global parental populations were assumed, but Figure 3 is based on six parental populations in Europe alone.

As noted earlier, this yields the somewhat strange idea that a population's "purity" depends on how closely the beholder looks at it. Perhaps when one has acne, close looks are not welcome, but in science they should be consistent with the interpretations being explicitly or implicitly given. Statistically, deviations from random mating among local subregions are small relative to the same differences in the context of a broader geographic sampling, as we would expect. However, either a population is "pure" in this sense or it is not, and such scale-dependence of the structure of populations is not exactly what people have in mind about "races" or types.

Now, if the measure used to define races is based on globally shared and, hence, relatively common allelic polymorphisms as is usually performed, then what we have is that a race is a polygenic statistical population. Indeed, globally common SNPs are perforce ancient and antedate the distinct types they are used to define. Further, a statistical definition implies that races are only quantitatively rather than qualitatively different, that is, they are not actual "types" because as mentioned earlier their definition allows each multisite genotype potentially to be found in any race.

This leads to a conundrum because if only nonshared variants were studied instead, one might think that groups could be identified once a sample was taken and thus could be used to define each race tautologically as a distinct type; it is a distinct type because it has distinct alleles. However, most nonshared variants will only be found in some individuals in a purported race's geographic homeland. That might seem to imply rather strangely that, depending on how definitions are operationalized, members of a purported race would not have all of its defining alleles! This shows how what one chooses to sample can affect or even predetermine the results. That is not supposed to happen in science.

Despite or, often we think, oblivious to these issues, the admixture-structure approach is nonetheless now routinely used in anthropological genetics, as exemplified by Figures 2 and 3 (Weiss and Long 2009; Weiss 2010; Weiss and Lambert 2010). Investigators are usually careful not to use words like "race," perhaps for political correctness, but the groupings are similar to classical races, and the ideas are the same in terms of the analysis used, disclaimers notwithstanding. This is clearly what one would expect of geographic samples of distantly located populations when the true generating process was basically an expansion for a human source population in Africa by gradual expansion of the leading edge of human population northward and eastward into Europe and Asia, a process that is not "admixture" between people from internally homogeneous, much less distant populations.

The availability of convenient statistical programs does not justify science that is knowingly misleading.

From an evolutionary point of view, what admixture analysis does is essentially to identify the historically and topographically induced irregularities in the otherwise roughly gradual pattern of change in genetic variation over geographic space, and it recognizes the increasing differences in peoples living farther apart. That is a true reflection of history as a process, except to the extent that it is colored, knowingly or otherwise, by selective de facto typological sampling and the assumption of statistically homogeneous source populations.

We can see the basic problem by reference to Figure 4, which imposes concepts of admixture seen in Figure 1 as they are now routinely used in structure analysis on the landscape produced by actual history. This shows what happens if a counterfactual assumption of admixture (as if it were like the lower panel of Fig. 1) is imposed on the reality panel (top): the dotted circles show how samples chosen from different parts of this actual continuity and treated *as if* they were isolated discrete or closed parental populations, who donated to samples from in between them. The result will be seen as an admixture result, simply because it is *assumed* to be that way.

SIMULATION ILLUSTRATES THE POINTS

One way to illustrate these points is to use a computer to simulate populations and their history in a reasonably realistic way, and then to examine the results as they are, and as they might appear if one made the kinds of admixture assumptions that we have been discussing. We have performed this with ForSim, a program

that we have developed for such purposes (Lambert et al. 2008; Weiss 2010; Weiss and Lambert 2010; or see popmodels.cancercontrol.cancer.gov/gsr/home). We simulated an initial population of size 1000 that expanded to 10,000 individuals during a run of 10,000 generations with 10 widely separated regions of DNA sequence each 30,000-nucleotides long (the spacing was performed to minimize correlations of sequence patterns among the 10 regions to make their variation statistically independent). Standard human mutation rates within and recombination rates between regions were applied to each generation. We specified that mutations arising in five of the simulated genes would add mutation-specific effects to the value of a simulated quantitative trait. These various values are consistent with estimates for the important basic parameters of the human population since its origin as a distinct species.

The simulated population expanded by diffusion from a local area outward across a square space represented by an X-Y coordinate grid. Each individual has a coordinate location at birth. Mating is random with respect to genotype and phenotype, but males choose females randomly from the surrounding adjacent coordinate locations. Offspring "live" in a location surrounding that of their "father" by a distance randomly chosen from a Normal (0,1) distribution of displacement in both X and Y directions. This is not intended as a rigorous but reasonable algorithm for simulating typical ancestral human mating distances and the diffusion of genetic variants (Cavalli-Sforza and Edwards 1964; Cavalli-Sforza et al. 1994). The values can all be adjusted in performing the sim-

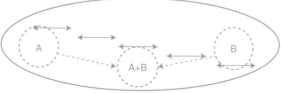

Figure 4. Admixture analysis can be imposed counterfactually on a continuous reality. Based on images from Figure 1, with, $K = 2$ parental populations, the *middle* population is analyzed as if it were a true admixed product of discrete populations A and B, which it is not (see text).

Cite this article as *Cold Spring Harb Perspect Biol* doi: 10.1101/cshperspect.a021238

ulation should one wish to explore the results. In some runs, we imposed weak directional selection on the trait to see whether that made any difference in the nature of the results.

At the end of the simulation, the simulated space was divided into 10 equal sampling boxes along the diagonal of the occupied space, and all of those boxes that had become inhabited during the run were used as "populations" for input to admixture-based analysis, with K set to the number of such samples. The results are shown in Figure 5 (in different runs, the number of populations ranged from $K = 7$ to 10). Note that imposing such boxes on what was, in fact, a continuously behaving distribution is just the kind of artifice that is typical of much of human genetics. Note that because the structure-analytic assumption is that one is not necessarily directly sampling any parental (they all are treated as ancestral), one can divide the result in arbitrary ways. Thus, population boundaries we used are not shown in the figure.

If one compares the typical empirical result shown in Figures 2 and 3 to the simulated data

in Figure 5, the general picture in the simulated data clearly resembles the real data. The resemblance is even greater for more geographically detailed real-data sets (e.g., Rosenberg et al. 2002; Li et al. 2008; Wang et al. 2008; Tishkoff et al. 2009). There is local similarity and more admixture between nearby regions. Yet, unlike real human history, the ForSim simulation involved no geographic irregularities or migration barriers that might affect the smooth diffusion of allele frequencies. Indeed, the structure analysis of the simulated data included all SNPs rather than using fewer, widely spaced, high-frequency SNPs likely to be polymorphic across the population range.

Figure 5 shows an inferred admixture history that, in the case of simulated data, is 100% fictional and reveals how thoroughly non-Darwinian are the assumptions of such analysis to real-world data. In fact, what we have simulated is essentially gradual isolation by the distance process of human population expansion and habitation, which is roughly how human history actually worked until the recent, rapid

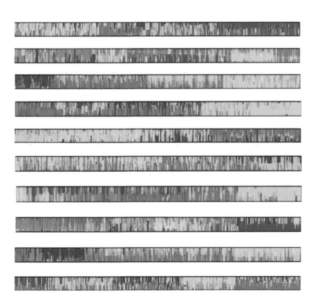

Figure 5. Admixture structure analysis when there is no admixture structure. As described in the text, results are shown from 10 independent ForSim simulations of gradual population expansion under identical conditions. In the admixture structure analysis of each run, $K = 7-10$ parental populations were assumed as program inputs (but not shown in the figure), although there were in fact no such discrete populations, nor any admixture among them. The data were analyzed using STRUCTURE (Pritchard et al. 2000), and results plotted with DISTRUCT (Rosenberg 2004), commonly used programs for this kind of analysis and portrayal.

large-scale distant-travel centuries. The differences among independent runs under the same parameters shown in Figure 5 also reveal the probabilistic (chance) aspects of what structure analysis will find as the appearance of multiple parental populations and their admixed descendants. Including natural selection in the simulations, geographic bottlenecks, a serial-founder expansion, or off-diagonal sampling boxes make no qualitative difference to the results (data not shown).

For Figure 5 we made no attempt to optimize the analysis because there are too many ways one might manipulate them to obtain desired results. But does the choice of K lead to an artifactual appearance of admixture structure in our simulated data? Figure 6 shows a comparison of one of our runs comparing $K = 8$ and $K = 3$ for data from a given simulated run; here, and in the published literature, these K values are arbitrary with regard to the qualitative nature of the resulting figures. Smaller K (fewer parental populations) generally yield an appearance of simpler, more "obvious" pure and admixed individuals. The results are essentially like those shown in Figure 2, which shows that what we are simulating reveals the empirical problems we have tried to raise. To be clear, the point of this type of simulation is not to generate a model of the real global human population, but to show that simulating the same kinds of processes as generated the geographic distribution of human genetic variation can give an entirely false appearance when analyzed under the counterfactual assumptions of structure analysis. In this sense, it is the assumptions of the analysis rather than true historical Darwinian facts that generate the results.

One might argue—correctly!—that the nature of these results is entirely obvious; if a program is designed to find ancestral populations and admixed individuals, then, of course, that is what it will find unless there was complete random mating in the entire global population. Nor, of course, does the similarity of simulated to real data by itself prove that admixture-based analysis is not finding historical truth. It works well, for example, when the situation is reasonably well understood historically, as in the case of African- or Mexican-Americans, whose admixture history was major, rapid, recent, and historically documented. But how can one tell? Because there are manifest reasons why there are not, and may never have been, truly isolated "ancestral" human populations in the admixture-structure sense, why use the approach or accept its results as true?

If this kind of analysis were strictly a digest of convenience as a way of portraying the relationship between location and genetic similarity, it could be unexceptionable. But investigators presenting such analysis almost universally imply real, not fictive history as if the parentals really did, or do, exist as such. Categorical treatment of humans has caused great grief in history, especially because if groups are different in terms of genetic variation and one takes a Darwinian assumption that traits are all here because of natural selection, then one easily slips into judgments of the inherent value of genes, and hence the people, in different race categories, whether that word is used to refer to the populations or not.

Instead of such an approach, there are other ways to analyze human variation, using the same data, that produce comparably esthetic graphical portrayal of its pattern over space that are more properly interpretable in terms of the actual evolutionary generating processes. Those processes have clearly been typified by

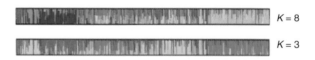

Figure 6. Assumptions affect results of admixture analysis. Smaller assumed numbers of ancestral populations make admixture from parental populations seem somewhat simpler because there are fewer parentals to draw from, although this is entirely an artifact of the analysis. Here, the same set of simulated data were analyzed as in Figure 5, assuming $K = 8$ (*top*) and $K = 3$ (*bottom*) parental populations.

local exchange of mates between nearby small populations as the human frontier gradually expanded out of Africa to populate the rest of the world. Even language, culture, and local geography as a rule only provide leaky barriers to such exchange. With today's instant access to global observations on a systematic scale, there should be no excuse for taking a typological viewpoint about human variation, be it statistical or otherwise.

WHY?

Why are we humans, scientists, and lay public alike so prone to perpetuate the categorical thinking of the past? What is at the root of such thinking? Is there any reason we should have come to that when we have always known that variation was more graded? Perhaps these are philosophical questions, or perhaps the answer would be a rather generic appeal to the evolutionary survival value of quick recognition of categories: food, mate, friend, or foe.

But this is the age of science in which we should be able to override such primeval reactions. The discovery of evolution as a population process that produces spatial and temporal variation in basically quantitative ways should have easily purged us of erroneous categorical thinking. There are, in fact, alternative ways to choose and analyze samples that do not make the typological assumptions.

These points are about admixture-based portrayals of human history. It may still be valuable to use structure-based analysis as a pragmatic way of accounting for uneven genotype frequency distributions across sampling space in the context of genetic mapping and inference related to identifying genes causally associated with diseases or other traits of interest. Even our pure diffusion-based simulated population process leads to genetic variation across space so that the spectrum of genotypes in one area is not identical to those in other areas; the greater the distance, the greater the difference.

This does not gainsay the value of some aspects of human categorical concepts. Self-defined ethnicity (Tang et al. 2005b) must be valuable as a convenient way to capture something about individuals sampled in epidemiology because it affects not just their marriage patterns, but also their habits and environmental exposures. That "something" is largely cultural, although it also has some correlation with geographic ancestry and, hence, genetic variation. Races, in this sense of social cohesion, that people choose to use about themselves and their community, certainly do exist; people with cultural affinities can have irregular and genome-wide characteristics, but whether or not they are usefully correlated with genetic causation of phenotypes such as disease, epidemiological objectives are different from inferring population history. History is a fact, not a convenience.

Linnaeus was a brilliant scientist who contributed foundational ways of organizing Nature's species. The Linnaean categorical framework raised the challenge to explain the origin of those categories and, perhaps, provided Darwin with a foil against which to evaluate his observations about variation. But one Linnaeus was enough.

ACKNOWLEDGMENTS

We thank the Editor for inviting us to contribute to this collection. We thank Anne Buchanan, and reviewers for helpful comments on this manuscript, and Noah Rosenberg for assistance with his program DISTRUCT. This does not imply their agreement with our point of view. Our work is supported by grant RO1 MH084995 from the National Institutes of Health, the Penn State Evan Pugh Professors Research Fund, and the Penn State Huck Institutes of the Life Sciences.

REFERENCES

Bauchet M, McEvoy B, Pearson LN, Quillen EE, Sarkisian T, Hovhannesyan K, Deka R, Bradley DG, Shriver MD. 2007. Measuring European population stratification with microarray genotype data. *Am J Hum Genet* **80:** 948–956.

Baur E, Fischer E, Lenz F. 1931. *Human heredity.* Macmillan, New York (first published in 1921).

Bryc K, Velez C, Karafet T, Moreno-Estrada A, Reynolds A, Auton A, Hammer M, Bustamante CD, Ostrer H. 2010. Colloquium paper: Genome-wide patterns of population structure and admixture among Hispanic/Latino populations. *Proc Natl Acad Sci* **107**: 8954–8961.

Carlson EA. 2001. *The unfit: A history of a bad idea.* Cold Spring Harbor Laboratory Press, Cold Spring Harbor, NY.

Cavalli-Sforza L, Edwards A. 1964. Analysis of human evolution. In *Proceedings of the 11th International Congress of Genetics*, Vol. 2, Pergamon, Oxford, pp. 923–933.

Cavalli-Sforza L, Menozzi P, Piazza A. 1994. *The history and geography of human genes.* Princeton University Press, Princeton, NJ.

Darwin CR. 1850. *Letter from Darwin, C.R. to Hooker, J.D. on 13 June [1850].* Cambridge University Library, Cambridge.

Hooton EA. 1926. Methods of racial analysis. *Science* **63**: 75–81.

Kevles DJ. 1995. *In the name of eugenics: Genetics and the uses of human heredity.* Harvard University Press, Cambridge, MA.

Kittles RA, Weiss KM. 2003. Race, ancestry, and genes: Implications for defining disease risk. *Annu Rev Genomics Hum Genet* **4**: 33–67.

Lambert BW, Terwilliger JD, Weiss KM. 2008. ForSim: A tool for exploring the genetic architecture of complex traits with controlled truth. *Bioinformatics* **24**: 1821–1822.

Li JZ, Absher DM, Tang H, Southwick AM, Casto AM, Ramachandran S, Cann HM, Barsh GS, Feldman M, Cavalli-Sforza LL, et al. 2008. Worldwide human relationships inferred from genome-wide patterns of variation. *Science* **319**: 1100–1104.

Lovejoy AO. 1936. *The great chain of being: A study of the history of an idea. The William James Lectures delivered at Harvard University, 1933.* Harvard University Press, Cambridge, MA.

Novembre J, Johnson T, Bryc K, Kutalik Z, Boyko AR, Auton A, Indap A, King KS, Bergmann S, Nelson MR, et al. 2008. Genes mirror geography within Europe. *Nature* **456**: 98–101.

Pritchard JK, Stephens M, Donnelly P. 2000. Inference of population structure using multilocus genotype data. *Genetics* **155**: 945–959.

Reich D, Thangaraj K, Patterson N, Price AL, Singh L. 2009. Reconstructing Indian population history. *Nature* **461**: 489–494.

Rosenberg NA. 2004. DISTRUCT: A program for the graphical display of population structure. *Mol Ecol Notes* **4**: 137–138.

Rosenberg NA, Pritchard JK, Weber JL, Cann HM, Kidd KK, Zhivotovsky LA, Feldman MW. 2002. Genetic structure of human populations. *Science* **298**: 2381–2385.

Schuchert C. 1897. What is a type in natural history. *Science* **5**: 636–640.

Shriver MD, Kittles RA. 2004. Genetic ancestry and the search for personalized genetic histories. *Nat Rev Genet* **5**: 611–618.

Tang H, Peng J, Wang P, Risch NJ. 2005a. Estimation of individual admixture: Analytical and study design considerations. *Genet Epidemiol* **28**: 289–301.

Tang H, Quertermous T, Rodriguez B, Kardia SL, Zhu X, Brown A, Pankow JS, Province MA, Hunt SC, Boerwinkle E, et al. 2005b. Genetic structure, self-identified race/ethnicity, and confounding in case-control association studies. *Am J Hum Genet* **76**: 268–275.

Tishkoff SA, Reed FA, Friedlaender FR, Ehret C, Ranciaro A, Froment A, Hirbo JB, Awomoyi AA, Bodo JM, Doumbo O, et al. 2009. The genetic structure and history of Africans and African Americans. *Science* **324**: 1035–1044.

Wagner J, Weiss KM. 2012. Attitudes on DNA ancestry tests. *Hum Genet* **131**: 41–56.

Wang S, Lewis CM, Jakobsson M, Ramachandran S, Ray N, Bedoya G, Rojas W, Parra MV, Molina JA, Gallo C, et al. 2007. Genetic variation and population structure in Native Americans. *PLoS Genet* **3**: e185.

Wang S, Ray N, Rojas W, Parra MV, Bedoya G, Gallo C, Poletti G, Mazzotti G, Hill K, Hurtado AM, et al. 2008. Geographic patterns of genome admixture in Latin American Mestizos. *PLoS Genet* **4**: e1000037.

Weiss KM. 2010. Does history matter? *Evol Anthropol* **19**: 92–97.

Weiss KM, Fullerton SM. 2006. Racing around, getting nowhere. *Evol Anthropol* **14**: 165–169.

Weiss KM, Lambert BW. 2010. When the time seems ripe. *Ann Hum Genet* **75**: 334–343.

Weiss KM, Long JC. 2009. Non-Darwinian estimation: My ancestors, my genes' ancestors. *Genome Res* **19**: 703–710.

 Cite this article as *Cold Spring Harb Perspect Biol* doi: 10.1101/cshperspect.a021238

Social Diversity in Humans: Implications and Hidden Consequences for Biological Research

Troy Duster

Chancellor's Professor & Senior Fellow, Warren Institute on Law and Social Policy, Boalt School of Law, University of California Berkeley, Berkeley, California 94720

Correspondence: troy_duster@berkeley.edu

Humans are both similar and diverse in such a vast number of dimensions that for human geneticists and social scientists to decide which of these dimensions is a worthy focus of empirical investigation is a formidable challenge. For geneticists, one vital question, of course, revolves around hypothesizing which kind of social diversity might illuminate genetic variation—and vice versa (i.e., what genetic variation illuminates human social diversity). For example, are there health outcomes that can be best explained by genetic variation—or for social scientists, are health outcomes mainly a function of the social diversity of lifestyles and social circumstances of a given population? Indeed, what is a "population," how is it bounded, and are those boundaries most appropriate or relevant for human genetic research, be they national borders, religious affiliation, ethnic or racial identification, or language group, to name but a few? For social scientists, the matter of what constitutes the relevant borders of a population is equally complex, and the answer is demarcated by the goal of the research project. Although race and caste are categories deployed in both human genetics and social science, the social meaning of race and caste as pathways to employment, health, or education demonstrably overwhelms the analytic and explanatory power of genetic markers of difference between human aggregates.

Two contradictory magnetic poles pull medical research on humans in opposite directions, producing a tension that will never be resolved. On the one hand, there is a universalizing impulse—based on a legitimate assumption that human bodies are sufficiently similar that vaccines, catheters, pasteurizing processes, and tranquilizers that work in one population will work in others. On the other hand, and unless and until research protocols establish and confirm specific similarities across populations, there is sufficient human variation that targeting medicines for specific populations can be a legitimate—even vital—empirically driven task. The theoretical question, of course, is why a particular population or subpopulation is to be so targeted? Because of folk theories about different groups' biological difference, or because of their social and political standing? Age, gender, and race leap to the forefront. The history of research on ailments as disparate as breast and prostate cancer (Rothenberg 1997; Wailoo 2011), heart disease (Cooper et al. 2005), and syphilis (Jones 1981; Reverby 2009) provides

Cite this article as *Cold Spring Harb Perspect Biol* doi: 10.1101/cshperspect.a008482

strong evidence that the answer is not either/or but *both*. So, on what grounds do we choose one strategy over the other?

And it is precisely on this point that Steven Epstein (2007) raises the most fundamental question:

> Out of all the ways by which people differ from one another, why should it be assumed that sex and gender, race and ethnicity, and age are the attributes of identity that are most medically meaningful? Why these markers of identity and not others? (Epstein 2007, p. 10)

The answer is profoundly social and political, economic, and cultural. The United States is the only country in the world that, as public health policy, does not operate on the assumption of the single standard human.

Moreover, by highlighting certain categories, there is the unassailable truth that other categories are thereby ignored. But more to the theoretical point, because each of the categories noted above has a potential or real biological base in either scientific or common sense understandings (Schutz 1962), when scientists report findings indicating differences, the danger is that these findings can seductively divert policymakers from seeking alternative interventions that could better address health disparities (Krieger 2011).

The goal of Epstein's monograph was to (a) better understand how ways of thinking about differences in human populations paved the way to try to "improve medical research by making it inclusive," and (b) explain how and why the strategies of inclusiveness got institutionalized:

> Academic researchers receiving federal funds, and pharmaceutical manufacturers hoping to win regulatory approval for their company's products, are now enjoined to include women, racial and ethnic minorities, children, and the elderly as research subjects in many forms of clinical research ... and question the presumption that findings derived from the study of any single group, such as middle-aged white men, might be generalized to other populations. (Epstein 2007, p. 5)

This shift has occurred only in the last two and a half decades, beginning with regulations that were developed first in 1986. Once again,

it is important to restate the relatively unique feature of this development as it applies mainly to the United States (Epstein 2007, p. 7). The rest of the world has continued to act on the presupposition of the standard human, at least until now. As we shall see, that is about to change.

THE GENOMIC REVOLUTION AND THE SEARCH FOR DIFFERENCES

At the end of the 20th century, the first draft of the Human Genome Map was completed, providing two kinds of hope for the near future. The first was quite explicitly about potential medical advances—that the completed map would spur the development of new kinds of therapies that would increase health and reduce the ravages of a wide variety of diseases. The second hope was more of a diffuse political aspiration, but it was loudly trumpeted at the famous White House news conference in June 2000. That was when President Bill Clinton (United States), Prime Minister Tony Blair (United Kingdom), and the two molecular geneticists who had led the public and private sector human genome projects all agreed that—citing findings from the first mapping and sequencing first draft— at the level of the DNA, there is no such thing as race.

However, regarding this pronouncement about the "end" of race, as Mark Twain once quipped about a newspaper article that reported that he had died, "the news of my death has been greatly exaggerated." So it has been with racial and ethnic categories. Indeed, there is substantial evidence that developments in several fields of inquiry relevant to molecular genetics (pharmacogenomics, pharmacotoxicology, clinical genetics, personalized medicine, and forensic science) have actually served to reinscribe race as a biological category (Duster 2005, 2006; Fullwiley 2007, 2008; Bolnick 2008; Kahn 2011; Roberts 2011).

Indeed, one of the most striking developments of the last few years has been the move by several governments to take strong protective "ownership" of the DNA of their own populations—a move designed to protect from possible biopiracy from the pharmaceutical industry in Western countries. Ruha Benjamin (2009) has called this "national genomic sovereignty," and it represents the opposite of the universal notion of human DNA envisaged at the completion of the Human Genome Project.

On the surface, this policy frame asserts a deeply nationalist sentiment of self-determination in a time of increasing globalization. It implicitly "brands" national populations as biologically distinct from other populations, "naturalizing" nation-state boundaries to ensure that less powerful countries receive the economic and medical benefits that may result from population genomics. (Benjamin 2009, p. 341)

Mexico amended its General Health Law in 2008 to make "the sampling of genetic material and its transport outside of Mexico without prior approval . . . illegal" (Séguin et al. 2008, p. 6).

The Genomic Sovereignty amendment states that Mexican-derived human genome data are the property of Mexico's government, and prohibits and penalizes its collection and utilization in research without prior government approval. It seeks to prevent other nations from analyzing Mexican genetic material, especially when results can be patented, and comes with a formidable bite in the form of prison time and lost wages. (Benjamin 2009, p. 344)

Mexico may be in the vanguard in so explicitly asserting its commitment to national "genomic sovereignty," but the nation is hardly alone. India, China, Thailand, and South Africa have all issued policy statements or passed legislation designed to develop national genomics infrastructure to benefit their populations (Séguin et al. 2008b).

In 2009, the HUGO Pan-Asian SNP Consortium, an international research team led by Edison Liu of the Genome Institute of Singapore, mapped genetic variation and migration patterns in 73 Asian populations, with data coming from 11 Asian countries: Japan, Korea, China, Taiwan, Singapore, Thailand, Indonesia, Philippines, Malaysia, Thailand, and India. The results—which included a summary statement that "there is substantial genetic proximity of SEA [Southeast Asian] and EA [East Asian] populations"—were published in the journal

Science (HUGO Pan-Asian Consortium 2009). In the same year, the Iressa Pan-Asian study (IPASS) was carried out by researchers in Hong Kong, mainland China, Thailand, Taiwan, and Japan with the participation of 87 centers in nine countries in Asia (Mok et al. 2009). This study was the result of previous research suggesting that Asian populations have a different, more positive response to this cancer drug than do other populations.

The explicit heightened racial consciousness of data reporting in human genomic science was dramatically on display in the November 6, 2008 issue of *Nature*. That journal published two articles asserting triumphantly how, for the first time, the whole human genome of (a) "an Asian individual," and then (b) of a Yoruban or "an African individual," were now "revealed" (Wang et al. 2008).

Nature referred to the fact that James D. Watson, Nobel Prize winner, as codiscoverer of the DNA structure, and J. Craig Venter, head of the private sector group that cosequenced the Human Genome, each has had their full genomes sequenced. Both are white males. But why this particular taxonomic system for trying to sort out useful, important, or relevant "differences?" The answer lies in a closer examination of recent emerging scientific discourse about "ancestral populations" and the fluid and contested boundaries around what constitutes a "population."

WHAT CONSTITUTES A "POPULATION" IN HUMAN GENETIC RESEARCH?

In 2007, *Science* magazine declared that genome-wide association studies (GWAS) were the scientific breakthrough of the year. Genome-wide association studies scan the genomes of large groups of individuals in search of markers that might be associated with specific common complex diseases (e.g., breast cancer). The frequency of a variant (single-nucleotide polymorphism or SNP) will differ across human populations.

Within a population, geneticists estimate the frequency at which a variant occurs in that population, based on a sample of individuals thought to belong to that population. (Fujimura and Rajagopalan 2011)

Which brings us to the key question: "What is a population?" A central task is to identify frequency differences between case and control groups that might be indicative of increased risk for the particular disease being studied. In the last decade, scores of research papers have been published emphasizing ethnic and racial differences between "populations." The term "populations" is in quotation marks for a compelling reason, namely, different researchers mean very different things when they use the term. From close observation of the laboratory work of geneticists who sample human groups across the globe, we now know that some use language group to mean a population, others take geographical boundaries; still others use already-collected data from previous research, in which it is unknown how the boundaries of the "population" in question were drawn, conceived, and implemented. In still other studies, a "population" is taken from the census; sometimes it is a "clinical population" as in those with a particular ailment—from cancer to hypertension, or from asthma to diabetes.

Yet another strategy is to find four grandparents whose ancestry can be traced to one of four broad continental groupings (Europe, Asia, Africa, the Americas). Although this may seem race neutral, on even superficial reflection, the social meaning of race is operating, because one does not mean ancestors who were Boers in South Africa, or grandparents who were European settlers in Quebec, or even great, great grandparents born in New Amsterdam.

This is but the tip of a numbing variety of factors that make human population strata and boundaries multilayered, porous, ephemeral, and difficult to identify (and thus):

Samples for genetic analysis are collected using operational criteria imposed by investigators and may be more representative of these operational criteria than actual breeding groups and gene pools. (Weiss and Long 2009, p. 704)

One of the most important tools now being deployed to examine human genetic variation is a computer-based program called STRUCTURE.

This program allows the researcher to identify patterns and/or clusters of DNA markers, and when an alignment of these clusters overlaps existing categories of race and ethnicity, there is the siren's seductive call to reinscribe these categories as biologically meaningful (Bolnick 2008). As I have suggested elsewhere, any computer program so instructed could find SNP pattern differences between randomly selected residents of Chicago and residents of Los Angeles (or between any two cities in the world). To put it in ways that are incontrovertible, no one could expect that SNP patterns would be identical in choosing subjects randomly from two cities. As for Chicago versus Los Angeles, such a proposed research project would be deemed ludicrous, because the theoretical warrant for it would be hard to establish (unless there was some legitimate grounding for hypotheses about smog effects vs. subzero winter effects). But if all the Chicago residents selected were African American, and all the Los Angeles residents were Asian American, and those SNP patterns showed up, an uncritical audience, lay or scientific, could easily accept these findings as having some validity affirming biological or genetic racial differences.

When is difference just difference, and when is difference something that inexorably stratifies a population? The answer lies in immediate history, context, and setting—in particular, whether there have been social meanings attributed to that differentiation. The authors of an often-cited piece in *Genome Biology* seem to acknowledge this when they say:

Finally, we believe that identifying genetic differences between racial and ethnic groups, be they for random genetic markers, genes that lead to disease susceptibility, or variation in drug response, is scientifically appropriate. What is not scientific is a value system attached to any such findings. Great abuse has occurred in the past with such notions as "genetic superiority" of one particular group over another. The notion of superiority is not scientific, only political, and can only be used for political purposes. (Risch et al. 2002, p. 11)

Although the sentiment is admirable, this formulation constitutes a fundamentally flawed notion of a firewall between "science" and "politics." All societies make sharp differentiations among their members that permit stratifying some groups over others. When humans create categories such as "caste" or "ethnic group" or "race," those taxonomies are political, and they are stratified in the most basic meaning of hierarchy: power-based differential access to resources. These three categories routinely predate and prefigure scientific inquiry, but, as I will demonstrate, profoundly constrain that inquiry. Over time, the interaction between living at the top or bottom of a stratified hierarchy produces systematized differential access to the rawest human needs. This means that there will be a feedback loop to various health and illness outcomes to those different "populations" (i.e., so stratified). If that seems abstract, here is a poignant example of that feedback loop.

Syngenta is one of the world's leading agribusiness companies, with more than 25,000 employees in nearly 100 countries across the globe. According to its official website, the company is dedicated to increase crop productivity through scientific advances, and to "protect the environment and improve health and quality of life." Syngenta has a plant in St. Gabriel, Louisiana, where it manufactures a crop-enhancing product called atrazine. But atrazine has an unfortunate side effect—it "demasculinizes and feminizes" vertebrate animals who are exposed to it by inducing aromatase.[1] When humans are exposed to atrazine for sustained periods, they are at a much-increased risk for certain cancers. The production facility in St. Gabriel has a prostate cancer rate 8.4 times higher among factory workers exposed to atrazine as compared with those in surrounding communities not exposed, and it just so happens that this plant is located in a community that is >80% African American (Hayes 2010, p. 3768).

These sharply different rates of prostate cancer between Whites and Blacks can be studied scientifically by geneticists trying to understand "population differences" through a unidimensional genetic prism, but with no understanding of the larger context in which humans are exposed to environmental insults—as in the first part of the formulation by Risch et al. But we can also study the systemic pattern of African Americans living close to toxic waste dumps across the whole country (Bullard 2000; Sze 2007). That is also available for systematic empirical investigation and testable formulations, otherwise known as science. Why should the decontextualized genetic inquiry of differing prostate cancer rates between Americans of European and recent African descent be characterized as apolitical "science," whereas the rate of their increased risk to exposure to atrazine is seen as "political" science? The answer is lodged in current culturally framed notions of the hierarchy of science. Being completely ahistorical and apolitical, we could take a sample of two different populations of Whites and Blacks in the contemporary United States, and we would find differences in their rates of hypertension. Although there is some debate about the extent of the gap, Blacks do tend to have somewhat higher rates than Whites. But as Richard Cooper and his colleagues (Cooper et al. 2005) have shown, by examining hypertension prevalence rates among 85,000 subjects, cross-cultural data demonstrate that this is not evidence for a biological difference between the races. It was explicitly designed to compare racial differences, sampling Whites from eight surveys completed in Europe, the United States, and Canada—and contrasting these results with those of a sample of three surveys among Blacks from Africa, the Caribbean, and the United States. The data from Brazil, Trinidad, and Cuba show a significantly smaller racial disparity in blood pressure than found in North America, and then, most tellingly, the authors of the study conclude:

These data demonstrate that the consistent emphasis given to the genetic elements of the racial contrasts may be a distraction from the more relevant issue of defining and intervening on the preventable causes of hypertension, which are likely to have a similar impact regardless of ethnic and racial background. (Cooper et al. 2005)

---

[1] Aromatase causes a higher estrogen-androgen ratio (Hayes 2010, p. 3768).

Yet the Cooper study, which involved more than 85,000 subjects across eight nations, was not taken as seriously as the study of 1056 African American subjects in a solely U.S.-based study of hypertension (Roberts 2011). Indeed, the FDA approved a drug designed by African Americans with hypertension the spring of the very same year, after the Cooper study was published (Cooper et al 2005; Kahn 2012). Because that decision was demonstrably more about economics, patenting, and politics than about science, it is naïve to think that these factors can be neatly parsed and isolated from each other.

The Transparent Conflation of Science and Politics: Genotyping Castes in India

In the introductory section, I noted that various nations around the globe have initiated their own genome projects, using national borders as the boundaries. When governments make these decisions, they are based on geopolitical considerations, not on human taxonomies generated by scientists. Nonetheless, when scientists then deploy these categories and boundaries, they are often reinscribing those very categories with scientific legitimacy and authority. It is imperative to address a fundamental misconception that when social scientists assert that some phenomenon (caste, class, race, ethnicity) is socially constructed (Haslanger 2008), the implication is that the phenomenon being examined is "not real." To make this point, take the example of money, or more specifically, the euro. It was obviously "socially constructed," because the German mark, the French franc, the Italian lira, and the Spanish peso (among other currencies) were converted into the euro by social, economic, and political forces and by a collaboration of decision makers. Having been thus "socially constructed," the euro is certainly "real." In a parallel manner, caste is socially constructed, but no less real in its consequences for life chances.

The differences in the way members of castes in India have systematically different outcomes regarding education, health, and economic well-being (or poverty level) is a direct consequence of social, cultural, economic, and

political forces. That would hardly be surprising, because endogamy rules (who can marry whom inside specific cultural categories) have precluded marriage (and to a lesser extent mating) practices for 30 centuries. Although these rules were never universally adhered to, they provided the frame in which dissent would be sometimes tolerated, sometimes sanctioned (Dirks 2001, p. 50; Kosambi 2002, p. 319). However, in a society riven by caste differences that persist to this day, what could it mean to demonstrate that there are discernible patterns of differences in the DNA of various castes (an outcome certainly to be expected after centuries of endogamy rules governing shaping marriage options)? Given the history that I am about to tell, any such differences discovered and reported regarding their respective DNA is a weapon in the hands of those who wish to explain and sustain their privilege. To wit, when some read about evidence that they have a somewhat distinctive set of microsatellite DNA markers, or DNA haplotypes, this can become grounds for suggesting that the differences in "education, health, and economic well-being (or poverty level) is a direct consequence" of genetic differences—not the other way around.

As a general phenomenon, elites of every society come to believe that their status, their high position in the social hierarchy, is both natural and just. Whether in caste, estate, or religious systems of stratification, those at the top are either universally born to privilege or frequently anointed at an early age. In class-based systems, those who themselves may have achieved a higher-class position by being mobile across class boundaries bequeath their status to their children. The oldest system of human stratification is in what in modern days we refer to as India. For much of India's history, the population has been divided into five major castes that do not intermarry and that have been forced into particular occupations by hereditary ascription. The top three castes are the Brahmins (priestly, literate), the Kshatriyas (mainly rulers and aristocrats), and the Vaisyas (businessmen). Together these three constitute ~17% of the population. The next group is the Sudras, who do the menial labor and are by far the largest

Cite this article as *Cold Spring Harb Perspect Biol* doi: 10.1101/cshperspect.a008482

varna, constituting about half of India's population. The last group are the Ati-Sudras, known by a variety of names, including Untouchables. Gandhi called them Harijans, or children of God, but they currently are most likely to go by the name of Dalits, or oppressed people.

The upper castes have excluded the lower castes from schools, post offices, restaurants, theaters, and barber shops. They were denied access to the courts, and, of course, with this record of exclusion, were never permitted to be employed in any professional occupation (Galanter 1984). As the priestly caste responsible for reading and interpreting the great books, the Brahmins had a monopoly on literacy. Given this history, it is hardly surprising that the Brahmins, <10% of the population, make up more than two-thirds of the students at the premier institution of higher education in the country, the University of New Delhi.

Although many Brahmins have been trumpeting the idea that India has ended the caste system and is celebrating individual meritocracy, the caste system lives on with fierce tenacity for much of India's 1.1 billion people. A series of studies on intergenerational mobility in India have produced the unsurprising finding that there is very little social mobility in the country (Dhesi and Sing 1989). Access to most good jobs is still restricted by longstanding cultural practices, and wage discrimination operates systematically for those from the Scheduled Castes (Lakshmanasamy and Madheswaran 1995).

That molecular geneticists find "allelic frequency differences" between castes should be no surprise. However, given the vast social, economic, and political gaps between castes, findings of "genetic differences" feed a newly molecularized interpretive account for those differences (health status, educational achievement, etc.). Even though these allelic frequency differences have no known function, reports of such findings constitute the basis for a molecular reinscription of caste differences. Here is some language capturing a crucial element of this trajectory:

We genotyped 132 Indian samples from 25 groups. . . . we sampled 15 states and six language families.

To compare traditionally "upper" and "lower" castes after controlling for geography, we focused on castes from two states: Uttar Pradesh and Andhra Pradesh.

We genotyped all samples on an Affymetrix 6.0 array, yielding data for 560,123 autosomal SNPs. . . . Allele frequency differentiation between groups was estimated with high accuracy (F_{ST} had an average standard error of 60.0011. . .).
(Reich et al. 2009)

Referring to an earlier study in which researchers found differing patterns of genetic markers between different socially designated groups, one Indian Genome Project coordinator explained that researchers "had intense debates on whether to reveal the names of communities. . . . I don't think scientists are prepared yet to understand the full social ramifications if such information is made public" (Mudur 2008).

Of course, Americans are prone to dismiss any parallels between caste domination in India and race privilege in the United States. It would hardly shock an American to learn that an imprisoned Brahmin had a better chance of employment in a decent paying job in corporate India than an Irula tribesman. Yet in the United States, the work of Devah Pager (2007) makes a powerful point that is a shocking parallel. In the last few years, Pager's research has become, quite deservedly, the poster child of a social science research project that is rigorous, critical, and saturated with both theoretical and policy implications. *The New York Times* heralded her research in a full-page report in March 2004, noting her finding that "it is easier for a white person with a felony conviction to get a job than for a black person whose record is clean" (Kroeger 2004).

This social meaning of race and caste as pathways to employment, health, or education demonstrably overwhelms the analytic and explanatory power of genetic markers of difference between human aggregates. However, there is a compelling reason why the Indian Genome Project coordinator (noted above) expressed concerns about "making public" data reporting genetic differences between castes. That concern is parallel to the molecular reinscription of race (Fullwiley 2007).

INSPECTING "POPULATION PURITY" IN HUMAN GENOMIC RESEARCH

In a recent paper that is yet another model of how social scientists can contribute to research in human molecular genetics, Hinterberger (2010) has looked at how Canadian scientists have been trying to better understand the biological sources of complex diseases by using GWAS. The assumption is that about 8500 French settlers arrived in Canada between 1608 and 1759. They intermarried among themselves and thus produced what is called a "founder effect":

> ... the Quebec "founder effect" has provided a large volume of genomic research aimed at understanding the root of common and complex disease. In 2007, a genome-wide association study...identified multiple genes underlying Crohn's disease in the Quebec founder populations (Raelson et al. 2007). GWAS are seen to offer a powerful method for identifying disease susceptibility for common diseases such as cancer and diabetes and are at the cutting edge of genomics-based biomedicine. (Hinterberger 2010, p. 15)

But now there is an explicit lament among these scientists who express concerns that intermarriage rates are threatening the "genetic uniqueness of these groups" and thus the opportunity for this kind of research "may be lost in the next few generations" (Secko 2008). Here is where Hinterberger (2010, 2012) steps in as the social analyst to point out the problematic unexamined assumptions of the presumed genetic homogeneity of the founding population of French settlers. What genetic researchers regard as a bounded French founder population is actually not so French after all. Specifically, in the strong pressure to convert indigenous people to Christianity, the colonizing French eagerly gave these converts French surnames (Kohli-Laven 2008). An examination of Parish records provides documentation of this, and yet it is the French name that demographers and historians have used to establish the assumption that those with French names constituted the *bounded* "French founder" population. This clearly upends the otherwise taken-for-granted assumptions about the homogeneity of this population. Without an appreciation of the social diversity

of the human population of founders, the geneticists' stated concerns about current interbreeding diluting the "pool" are misguided. This brings us to a discussion of similar concerns about the use of ancestry markers.

With very few exceptions, ancestry-informative markers are shared across all human groups. It is therefore not their presence or absence, but their rate of incidence, or frequency, that is being analyzed. When taken together, these markers appear to yield certain patterns in people and populations tested. A specific pattern of alleles on each of a set of chromosomes that have a high frequency in the "Native Americans" sampled then become established as a "Native American" ancestry signature. The problem is that millions of people around the globe will have a similar pattern, that is, they will share similar base-pair changes at the genomic points under scrutiny. This means that someone from Hungary whose ancestors go back to the 15th century could map as partly "Native American," although no direct ancestry is responsible for the shared genetic material. Ancestry-informative markers, however, arbitrarily reduce all such possibilities of shared genotypes to "inherited direct ancestry." In so doing, the process relies excessively on the idea of 100% purity, a condition that could never have existed in human populations.

To make claims about how a test subject's patterns of genetic variation map to continents of origin and to populations where particular genetic variants arose, the researchers need reference populations. The public needs to understand that these reference populations comprise relatively small groups of contemporary people. Moreover, researchers must make many untested assumptions in using these contemporary groups to stand in for populations from centuries ago, representing a continent or an ethnic or tribal group. To construct tractable mathematical models and computer programs, researchers make many assumptions about ancient migrations, reproductive practices, and the demographic effects of historical events such as plagues and famines. Furthermore, in many cases, genetic variants cannot distinguish among tribes or national groups because the groups are too similar, so geneticists are on thin ice when telling people that they do or do not have ancestors from a particular people.

Instead of asserting that someone has no Native American ancestry, the most truthful statement would be: "It is possible that although the Native American groups we sampled did not share your pattern of markers, others might, because these markers do not exclusively belong to any one group of our existing racial, ethnic, linguistic, or tribal typologies." But computer-generated data provide an appearance of precision that is dangerously seductive.

There is a yet more ominous and troubling element of the reliance on DNA analysis to determine who we are in terms of lineage, identity, and identification. The very technology that tells us what proportion of our ancestry can be linked, proportionately, to sub-Saharan Africa (ancestry-informative markers) is the same being offered to police stations around the country to "predict" or "estimate" whether the DNA left at a crime scene belongs to a white or black person. This "ethnic estimation" using DNA relies on a social definition of the phenotype. That is, to say that someone is 85% African, we must know who is 100% African. Any molecular, population, or behavioral geneticist is obliged to disclose that this "purity" is a statistical artifact that begins not with the DNA, but with a researcher's adopting the folk categories of race and ethnicity.

Sampling for Human Genetic Diversity — The Conundrum[2]

Researchers ideally would like to sample to achieve representativeness of diversity. But in order to "sample," one must have a notion of the boundaries of the larger population base from which one is sampling. Yet those boundaries are always going to be absent when it comes to human genetic diversity, because unless or until we have a Wilson-type grid for the world's population, we will not have a firm empirical basis for understanding who any "sample" represents. Short of an empirical basis for proceed-

ing with a sound sampling strategy, we are then left with this conundrum of talking about "sampling" when there is no bounded population delimited by some theoretical frame. Of course, that is where race and ethnicity tend to surface in these discussions, but there are bundles of unpacked assumptions built in to the idea that any five sets of people represent five races—whether biologically or socially!

There is yet a prior question of what is meant by "diversity," and on that matter, it is vital to be really clear on the substantive meaning of genetic diversity. If the goal is to capture genetic diversity, the strategy might aim to obtain samples from people who are presumably as genetically different from each other as possible. If that is the goal, then the researcher is simply trying to capture a wide range of specifiable variation. Here is where we must get to substance, because there are numerous dimensions and levels on which people can vary from each other genetically. Thus, the idea of a high degree of variation may not be meaningful because it is not likely to capture the type of genetic variation in which the researcher is most interested. Or conversely, such a strategy might capture a lot of variation in which the researcher has little interest, for example, variation of little apparent relevance to health outcomes.

Is the researcher trying to "represent" the range of genetic variation in a specific region? In this case, one does have a sampling problem, and unless one is assuming a level of homogeneity (that is impossible to demonstrate empirically), this cannot be done with a few dozen people. Yet the report from one of the early studies, although well-intentioned, well-crafted, and designed to help better understand health differences in variable human population groups (Hinds et al. 2005), does point in that direction. The researchers were searching for, and found, patterns of SNPs differentially distributed in three population groups, formed from a total of 71 persons who were either Americans of African descent, Americans of European descent, or Han Chinese.

The title of the paper is instructive, "Whole Genome Patterns of Common DNA Variation in Three Diverse Human Populations." Howev-

[2]The following discussion is indebted to Pilar Ossorio.

er, what makes these three populations diverse is the phenotype associated with a racial classification system, not a genotypic pattern of similarity that triggered the inquiry. Indeed, the authors note that the SNP patterns of genetic diversity that they found among African Americans suggest a more substantial diversity than that in the other two populations, a finding consistent with our knowledge of genetic diversity on the African continent. So, why was the question raised in this manner? The answer is a scientific catch-22. The main reason is convenience: The data were collected and marked that way in the Coriell Cell Repositories. That is an understandable rationale. However, by deploying these existing categories, any differences that emerge are likely to be "racialized," no matter how many caveats and demurrers appear in the text of a scientific paper. Moreover, the African American group is said to be "admixed." But in terms of the genotype, all three groups are "admixed." So it must be the phenotype to which the authors refer with the designation of "three diverse populations."

The clinical manifestation of a health problem might be primarily a consequence of social, cultural, and economic forces and might have little to do with "genomic diversity." For instance, as noted above, living near a toxic waste site may increase one's chances for developing cancer or asthma, or differing nutritional intake patterns may produce diabetes, obesity, asthma, or hypertension. Some social, cultural, and economic factors influence epigenetic processes and gene expression. Researchers will want to thoroughly characterize the iPS cells, including not only DNA sequence variation but epigenetic markings, expression profiling, metabolic profiling, etc. Socioeconomic variation that influences epigenetics or other biological phenomena may also be important to sample. Thus, diversity criteria might also include factors such as immigrant/nonimmigrant status, wealth level, educational level, and other social factors known to influence health outcomes. Those nongenetic criteria help take the focus off of race as if it were primarily a biological variable and sharply reduce attendant concerns about sampling for genetic diversity.

CONCLUSION

We return to the difficult and vexing question with which we began but for which some answers are now available. Of all the myriad ways in which humans differ, why are some categories chosen (and others ignored) in order to map human diversity for the purposes of population-specific treatment regimens, pharmaceuticals, vaccines, and even patterns of migration across the globe? The most compelling answer begins with an acknowledgement of the social aspects of the phenotype. Caste is not a biological category, it is a social category. However, when human molecular geneticists sort "populations" by these social categories and find (inevitably) different patterns of the frequency of genetic markers in those very social categories, the larger social context of those findings are arrestingly seductive as a framework for explaining differential life chances outcomes. This process constitutes the "molecularization" (Rose 2001, 2007; Fullwiley 2008) and the "geneticization" (Lippman 1991) of explanations of complex social forces.

The social, economic, and legal consequences evolve into several dimensions and have now reverberated into how patents to biotech companies are granted. Jonathan Kahn has documented how this process has unfolded at the United States Patent and Trademark Office:

> ... the practice of requiring patent applicants to introduce race into their biotech patents has become routinized at the USPTO. (Kahn 2011, p. 402)

So what begins with the assumption that researchers are pursuing neutral, apolitical science when they deploy folk categories of caste, race, or ethnicity, can seamlessly segue into reified practices that deliver targeted pharmaceuticals to racialized target populations (and we can assume, "caste-ized" target populations, following the Indian Genome Project) under the banner of personalized medicine. This train has already left the station (Tayo et al. 2011), and all that is left to determine is which track it will be on. A closer monitoring of the hidden assumptions will at least avoid some unfortunate collisions with the social realities (of the social diversity) of human populations.

Cite this article as *Cold Spring Harb Perspect Biol* doi: 10.1101/cshperspect.a008482

REFERENCES

Benjamin R. 2009. A lab of their own: Genomic sovereignty as postcolonial science policy. *Policy Soc* **28**: 341–355.

Bolnick D. 2008. Individual ancestry inference and the reification of race as a biological phenomenon. In *Revisiting race in a genomic age* (ed. Koenig BA, Lee SS-J, Richardson SS), pp. 70–85. Rutgers University Press, New Brunswick, NJ.

Bullard RD. 2000. *Dumping in Dixie: Race, class, and environmental quality.* Westview Press, Boulder, CO.

Cooper RS, Wolf-Maier K, Luke A, Adeyemo A, Banegas JR, Forrester TE, Giampaoli S, Joffres M, Kastarinen M, Primatesta P, et al. 2005. An international comparison study of blood pressure in populations of European vs. African descent. *BioMed Central* **3**: 1–8.

Deshi AS, Singh S. 1989. Education, labor market distortions and relative earnings of different religion-caste categories in India. *Can J Dev Studies* **10**: 75–89.

Dirks NB. 2001. *Castes of mind: Colonialism and the making of modern India.* Princeton University Press, Oxford.

Duster T. 2005. Race and reification in science. *Science* **307**: 1050–1051.

Duster T. 2006. The molecular reinscription of race. *Patterns Prejudice* **40**: 427–441.

Epstein S. 2007. *Inclusion: The politics of difference in medical research.* University of Chicago Press, Chicago.

Fujimura JH, Rajagopalan R. 2011. Different differences: The use of 'genetic ancestry' versus race in biomedical human genetic research. *Soc Stud Sci* **41**: 5–30.

Fullwiley D. 2007. The molecularization of race: Institutionalizing human difference in pharmacogenetics practice. *Sci Cult* **16**: 1–30.

Fullwiley D. 2008. The biologistical construction of race: "Admixture" technology and the new genetic medicine. *Soc Stud Sci* **38**: 695–735.

Galanter M. 1984. *Competing equalities: Law and the backward classes of India.* University of California Press, Berkeley, CA.

Haslanger S. 2008. A social constructionist analysis of race. In *Revisiting race in a genomic age* (ed. Koenig BA, Lee SS, Richardson SS), pp. 56–69. Rutgers University Press, New Brunswick, NJ.

Hayes TB. 2010. Diversifying the biological sciences: Past efforts and future challenges. *Mol Biol Cell* **21**: 3767–3769.

Hinds DA, Stuve LL, Nilsen GB, Halperin E, Eskin E, Ballinger DG, Frazer KA, Cox DR. 2005. Whole-genome patterns of common DNA variation in three human populations. *Science* **307**: 1072–1079.

Hinterberger A. 2010. The genomics of difference and the politics of race in Canada. In *What's the use of race: Modern governance and the biology of difference* (ed. Whitmarsh I, Jones DS). MIT Press, Cambridge, MA.

Hinterberger A. 2012. Categorization, census, and multiculturalism: Molecular politics and the material of nations. In *Genetics and the unsettled past: The collision of DNA, race, and history* (ed. Wailoo K, Nelson A, Lee C). Rutgers University Press, New Brunswick, NJ.

Jones JH. 1981. *Bad blood: The Tuskegee syphilis experiment.* Free Press, New York.

Kahn J. 2011. Mandating race: How the USPTO is forcing race into biotech patents. *Nature Biotechnology* **29**: 401–403.

Kahn J. 2012. *Race in a bottle: The story of BiDil and racialized medicine in a post-genomic age.* Columbia University Press, New York.

Kohli-Laven N. 2008. Hidden history: Race and ethics at the peripheries of medical genetic research. *GeneWatch* **20**: 5–7.

Kosambi DD. 2002. *An introduction to the study of Indian history.* Popular Prakashan, Mumbai, India.

Krieger N. 2011. *Epidemiology and the people's health.* Oxford University Press, New York.

Kroeger B. 2004. When a dissertation makes a difference. *The New York Times*, March 20.

Lakshmanasamy T, Madheswaran S. 1995. Discrimination by community: Evidence from Indian scientific and technical labor market. *Indian J Soc Sci* **8**: 59–77.

Lippman AJ. 1991. Prenatal genetic testing and screening: Constructing needs and reinforcing inequalities. *Am J Law Med* **17**: 15–50.

Mok TA, Yi-hong W, Thongprasert S, Chih-Chih Y. 2009. Gefitinib or Carboplatin-Paclitaxel in pulmonary adenocarcinoma. *N Engl J Med* **361**: 947–957.

Mudur GS. 2008. Stamp on Tagore's India genetic map blurs lines. *The Telegraph, Calcutta, India*, April 25.

Pager D. 2007. *Marked: Race, crime, and finding work in an era of mass incarceration.* University of Chicago Press, Chicago.

Raelson JV, Little RD, Ruether A, Fournier H, Paquin B, Van Eerdewegh P, Bradley WEC, Croteau P, Nguyen-Huu Q, Segal J, et al. 2007. Genome-wide association study for Crohn's disease in the Quebec founder population identifies multiple validated disease loci. *Proc Natl Acad Sci* **104**: 14747–14752.

Reich DK, Thangaraj N, Patterson N, Price AL, Singh L. 2009. Reconstructing Indian population history. *Nature* **461**: 489–495.

Risch N, Burchard E, Ziv E, Tang H. 2002. Categorizations of humans in biological research: Genes, race and disease. *Genome Biol* **3**: comment2007.1–comment2007.12.

Reverby S. 2009. *Examining Tuskegee: The infamous syphilis experiment and its legacy.* University of North Carolina Press, Chapel Hill, NC.

Roberts D. 2011. *Fatal invention: How science, politics, and big business re-create race in the twenty-first century.* The New Press, New York.

Rose N. 2001. The politics of life itself. *Theory, Cult Soc* **18**: 1–30.

Rose N. 2007. *The politics of life itself: Biomedicine, power and subjectivity in the twenty-first century.* Princeton University Press, Princeton, NJ.

Rothenberg K. 1997. Breast cancer, the genetic "quick fix" and the Jewish community: Ethical, legal and social challenges. *Health Matrix* **7**: 97–124.

Schütz A, 1962. Common sense and scientific understandings. In *Collected works of Alfred Schütz*, Vol. 1. Martinus Nijhoff, The Hague, Netherlands.

Secko D. 2008. Rare history, common disease. *The Scientist* **22** (July): 38.

Séguin B, Hardy B, Singer PA, Daar AS. 2008. Genomics, public health and developing countries: The case of the Mexican National Institute of Genomic Medicine (IN-MEGEN). *Nat Rev Genet* **9:** S5–S9.

Sze J. 2007. *Noxious New York: The racial politics of urban health and environmental justice.* MIT Press, Cambridge, MA.

Tayo BO, Tell M, Tong L, Quin H, Khitrov G, Zhang W, Song Q, Gottesman O, Zhu X, Pereira Ac, et al. 2011. Genetic background of patients from a university medical center in Manhattan: Implications for personalized medicine. *PLoS ONE* **6:** e19166.

Wailoo K. 2011. *How cancer crossed the color line.* Oxford University Press, New York.

Wang J, Wang W, Li R, Li Y, Tian G, Goodman L, Fan W, Zhang J, Li J, Xhang J, et al. 2008. The diploid sequence of an Asian individual. *Nature* **56:** 60–65.

Weiss KM, Long JC. 2009. Non-Darwinian estimation: My ancestors, my genes' ancestors. *Genome Res* **19:** 703–710.

Cite this article as *Cold Spring Harb Perspect Biol* doi: 10.1101/cshperspect.a008482

Demographic Events and Evolutionary Forces Shaping European Genetic Diversity

Krishna R. Veeramah[1] and John Novembre[2]

[1]Arizona Research Laboratories Division of Biotechnology, University of Arizona, Tucson, Arizona 85721

[2]Department of Human Genetics, University of Chicago, Chicago, Illinois 60637

Correspondence: jnovembre@uchicago.edu

Europeans have been the focus of some of the largest studies of genetic diversity in any species to date. Recent genome-wide data have reinforced the hypothesis that present-day European genetic diversity is strongly correlated with geography. The remaining challenge now is to understand more precisely how patterns of diversity in Europe reflect ancient demographic events such as postglacial expansions or the spread of farming. It is likely that recent advances in paleogenetics will give us some of these answers. There has also been progress in identifying specific segments of European genomes that reflect adaptations to selective pressures from the physical environment, disease, and dietary shifts. A growing understanding of how modern European genetic diversity has been shaped by demographic and evolutionary forces is not only of basic historical and anthropological interest but also aids genetic studies of disease.

During classical antiquity, writers such as Herodotus chronicled the expansion and contraction of empires, as well as the traditions of the peoples associated with them. Julius Caesar's memoirs from the Roman conquests of Gaul detail his encounters with foreign tribes such as the Helvetii and the Belgae. Such accounts were fascinating to peoples of that era as humans lived largely in ignorance of other cultures beyond their relatively small geographical vicinity. In the modern world, the barriers to acquiring knowledge of other contemporary societies are small; we can now easily learn about populations from across the world through an abundance of sources. Instead, the major challenge is to discern whom the peoples of the past were. From the perspective of genetics, we are especially curious about how past demographic and evolutionary events influenced the genetic diversity in humans today. However, peering into the past poses major challenges, and, in some ways, we stand much like Herodotus and Caesar, trying to piece together an understanding of distant populations from limited contact and partial experiences.

For geneticists, Europe represents a uniquely well-studied region of the world. On the one hand, it has a richness of accessible sources. We have already mentioned historical accounts beginning with the ancient works of Herodotus; there has also been an abundance of archaeological, anthropological, and linguistic studies.

More recently there has been substantial interest in understanding the genetic history of modern Europeans. Indeed, many of the largest studies of the genetics of human populations have taken place in Europe. This is, in large part, because of the availability of European universities and biomedical centers, which have provided the infrastructure for such "big science" studies that other regions have traditionally lacked. On the other hand, European human diversity has at various times been highly politicized, which has led to deeply misguided perspectives on the subject of genetic superiority and some of the most atrocious abuses to human life—the genocides and eugenics of the first half of the 20th century.

Contemporary genetic studies in Europe still work under the shadow of such views that are now understood as being scientifically without merit as well as ethically wrong, and, as today's scientists, we must be sensitive to the potential future misuse of findings regarding genetic diversity. That said, the field has been reinvigorated during the past approximately 50 years as perspectives on human diversity, both cultural and genetic, have matured. Scientifically, it is now appreciated that the genetic differences among humans are, in absolute terms, small as first identified by Lewontin (1972) (also see Chakravati 2014). Simplistic notions of genetic determinism have also fallen aside as most human traits are now thought to be driven by complex interactions between multiple environmental and genetic factors. Culturally, there is a wider appreciation that diversity makes a positive contribution to society. And finally, it is now recognized that understanding background patterns of genetic diversity is an essential component for combating heritable and infectious diseases.

Thanks to the growing interest in human population genetics, the scale of recent studies of European genetic diversity has grown to a staggering extent. Studies involving Europeans are some of the largest to have been performed in any population, regardless of the species. As a result, research on genetic diversity in Europe is of interest not just to scientists examining other human populations around the world but to all students of genetic diversity.

MAJOR HYPOTHESIZED DEMOGRAPHIC FACTORS SHAPING GENETIC DIVERSITY IN EUROPE

In a landmark review, Barbujani and Goldstein (2004) highlighted three major prehistoric demographic events that are likely to have had an impact on European genetic diversity (Fig. 1). The first event they described was the initial colonization of Europe by hunter–gatherers approximately 40,000 years ago. These travelers, likely consisting of only one or a few small groups that split off from established populations in northeast Africa near the Levant, therefore, represented only a small subset of the total human genetic diversity present within Africa. The entryway for humans into Europe is thought to have been via the Near East through modern-day Turkey, after which they expanded to the northwest either using a direct route or by first traveling along the coasts of the Mediterranean before turning northward. One plausible genetic signature of this process would be the reduction of genetic diversity as one moves northward and westward in Europe.

Second, Barbujani and Goldstein highlighted the constriction of northern European populations southward caused by the expansion of ice sheets during the last glacial maximum (LGM) at the end of the Pleistocene era (approximately 18,000 years ago). Evidence for the impact of Pleistocene glacial movements on genetic diversity has been observed in studies of a myriad of animal and plant species (Hewitt 1999). It seems unlikely that humans would have been exempt from the effects of these dramatic climatic shifts. The genetic signatures that may characterize this period may be differentiation among modern European populations reflecting the geographic locations of the different ancient glacial refugia, as well as less genetic diversity in northern Europe because of larger, more stable population sizes in the south and postglacial recolonization northward (Fig. 1).

The third possible event noted by Barbujani and Goldstein (2004) was the expansion of the first farmers into Europe following the emergence of agriculture in the Near East approxi-

Cite this article as *Cold Spring Harb Perspect Biol* doi: 10.1101/cshperspect.a008516

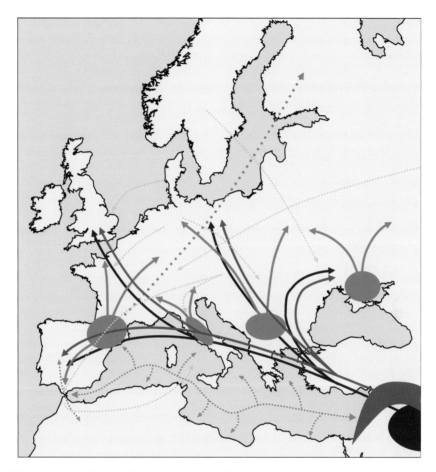

Figure 1. Schematic map of Europe showing the major migrations events that may have influenced modern-day European genetic diversity. The map shows (1) the initial colonization of Europe approximately 40,000 years ago from the Middle East (blue arrows), (2) the contraction of humans into four major refugia during the LGM approximately 18,000 years ago followed by the subsequent recolonization of Europe (green arrows), (3) the movement of Neolithic famers approximately 10,000 years ago from the Fertile Crescent (red arrows), (4) various barbarian migrations into the Roman Empire during the migration period of ~400–800 CE (dashed yellow arrows), and (5) continued gene flow across the Mediterranean Sea and between the Iberian Peninsula and North Africa throughout human history (dashed orange arrows). (This map is adapted from that presented in Renfrew 2010, which was originally made by Alessandro Achilli and Antonio Torroni.)

mately 10,000 years ago, which signified the dawn of the Neolithic era. The spread of agriculture observed in the archaeological record may have been strictly cultural, but many researchers have argued that this was accompanied by a demographic expansion in which incoming agricultural communities displaced hunter–gatherer groups. Genetic signatures of this "demic diffusion" hypothesis would be a gradient in gene (allele) frequencies from the Near East toward Western Europe, as well as

that of mixing between immigrant Neolithic and resident Paleolithic populations.

At the time of the Barbujani and Goldstein (2004) review, there was little evidence for the importance of interbreeding between modern humans emerging out of Africa and closely related forms of archaic humans, such as Neanderthals, who were already established across much of Europe. However, the sequencing of a Neanderthal genome (Green et al. 2010) has since provided evidence that up to 4% of current non-African human genetic ancestry (see Box 1) is from Neanderthal sources. However, whether there was admixture specifically with Paleolithic Europeans or whether the Neanderthal ancestry found in modern Europeans derives from gene flow that occurred in the ancestors of all non-Africans shortly after they left Africa is still largely unknown. It will be interesting to see if more detailed studies in the future can shed further light on this question. For example, there is emerging evidence of a distinct period of gene flow with the ancestors of contemporary Asians (Wall et al. 2013).

Beyond these events, three other demographic factors also likely have played at least some role in shaping modern European genetic diversity. One is what historians refer to as the migration period of Europe from ∼400–800 CE (Common Era). Historical records suggest that there was an extensive invasion of the Roman Empire by barbarian tribes such as Goths, Lombards, and Slavs, and many modern-day nation states trace their identity to these peoples. However, historians such as Patrick Geary have called into question whether mass movements of large cultural units actually took place (Geary 2003). Instead, he suggests that the ethnogenesis of many groups represented the influence of a military or aristocratic elite that

opportunistically drew adherents from local populations; thus, there would have likely been very little impact on modern genetic diversity.

Another factor is the contact of European peoples with those from neighboring geographical regions. Although the borders of Europe are generally considered the Mediterranean Sea and Caucasus Mountains, these are not strict impediments to gene flow. Admixture between North Africans and Europeans almost certainly occurred during the Moorish occupation of the Iberian Peninsula, and there is an age-old complex system of commerce across the Mediterranean. Likewise, contacts with western Asia and the Middle East via trade, or via more ancient contacts, likely also contributed to European genetic diversity.

Finally, among all of these major migrations, one must not lose sight of the behavior of the typical individual. Like most humans from across the world, the average European would likely find a mate locally. In some cultures, an average male may have moved less, as marriage practices may be distinct between the genders and show a patrilocal pattern in which wives move further from their birthplace by relocating to the birth location of their husbands. The long-term consequences of this local, sex-biased dispersal to a specific region

BOX 1. GENETIC ANCESTRY

Throughout the article, we often refer to the concept of the "genetic ancestry" of an individual. Although it is a somewhat ambiguous and often loosely applied term, scientifically, genetic ancestry is based on the idea that any particular individual is part of a large human pedigree and related to certain ancestors in this pedigree moving back in time (i.e., two parents, four grandparents, eight great grandparents, etc.). In theory, the number of ancestors increases exponentially, but in reality there can only be a finite number for all persons. When genetic material is passed from an ancestor to a descendent in this pedigree, a process called recombination reshuffles the two chromosomes a particular individual inherited from their two parents, after which this new mix of chromosomes is passed on to a subsequent child. The consequence of this process is that any individual's genome is made up of a mosaic of genetic material from many of their ancestors back in time. If we take an arbitrary point back in time, some of these ancestors may derive from a certain population, whereas the remainder may derive from another genetically distinct population. For example, for an African American individual, 75% of their genome may derive from ancestors from sub-Saharan African, whereas 25% of their genome may have been inherited from European individuals. Similarly, a modern-day European individual could possess genetic ancestry from both northwestern and southeastern European ancestors.

Cite this article as *Cold Spring Harb Perspect Biol* doi: 10.1101/cshperspect.a008516

would be a genetic similarity among distinct geographic neighbors, as well as differences in genetic material inherited through male and female lines (i.e., the Y chromosome and mitochondrial DNA).

How the various demographic factors described above have meshed to create the current diversity of modern Europeans is a fascinating yet complicated question to answer. For instance, our summary has highlighted several events that all predict a higher degree of genetic diversity in southern than northern Europe, making it difficult to identify which specific events truly took place as the genetic signatures of the events would overlap. This situation has eloquently been compared with a "palimpsest," an ancient manuscript that has been partially cleaned for novel writing, such that traces of the original writing remain (Jobling et al. 2003). Therefore, we should recognize that it is simplistic to think that genetic signatures of these distinct events can be clearly delineated. With these complexities in mind, we now turn to what has been learned about patterns of human genetic diversity from contemporary Europeans.

GENES AND GEOGRAPHY IN EUROPE VIA GENOME-WIDE DATA

In summarizing the patterns of diversity in European genetic data, we will take a personal perspective based on some of the studies we have been a part of, and then we will link out to studies that have preceded and followed our own work. In 2005–2006, DNA microarrays (a technology that uses a large number of DNA probes arrayed at microscopic scale on a solid surface) began to emerge as an efficient technology to determine which genetic sequence variants an individual carried (their genotype) at hundreds of thousands of individual locations in their genome in a single experiment. The costs became low enough that these arrays could be tested on thousands of individual samples, opening the way for large-scale population-based genetic studies that attempted to link disease susceptibility to particular genomic regions. As these population-based studies were

being performed, a major concern was whether subtle mismatches in genetic ancestry between disease cases and disease-free or random controls would lead to false-positive disease-causing alleles being implicated because of natural allele frequency differences between these two groups. To help avoid this, one might try to match every case individual with a matched control of the same genetic ancestry. One effort was to produce a large reference data set of healthy controls: the POPRES study (Nelson et al. 2008). The European subsample of the POPRES data set contained genotypes at 500,568 locations in the genomes of 3192 individuals from 37 different populations (which mostly represented countries of birth for the individual or their grandparents).

One of the major patterns to immediately emerge from a study on this data by one of us (J.N.) was that genetic relationships among individuals were strongly correlated with their geography, such that a visual summary of the genetic data looked strikingly similar to the geographic map of Europe (Fig. 2) (Novembre et al. 2008). Notably, we did not see discrete subgroups of individuals, which might arise, for instance, if a major factor affecting modern European genetic diversity had been its subdivision into separate glacial refugia or if major language groups (e.g., Romance, Slavic, Germanic) had created substantial long-term barriers to local mating. The only major genetic outliers were five Italian samples, which, based on later analyses, we now suspect are from Sardinia, an isolated island population in the Mediterranean (see below). At roughly the same time, other groups were finding similar results (Heath et al. 2008; Lao et al. 2008). The overall pattern to emerge from these studies of Europe is of a subtle genetic continuum among modern Europeans that is determined by geography. Recent work by Ralph and Coop (2013) suggests that much of this signal can be observed even while focusing on recent common ancestry, and that the most geographically separated Europeans have shared hundreds of common ancestors within the last 3000 years.

From this broad view of genetic variation in Europe, it is not clear what impact the many

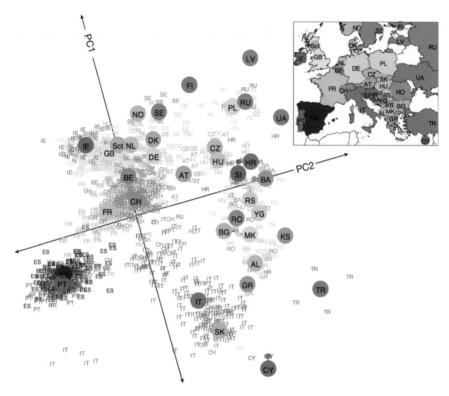

Figure 2. A statistical summary of genetic data from 1387 Europeans based on the two components that explain the greatest variation in the data. Small colored labels represent individuals and large colored points represent median coordinate values for each country. The *inset* map provides a key to the labels. The axes are rotated to emphasize the similarity to the geographic map of Europe.

historical and demographic factors alluded to at the beginning of this article have had on their genomes. If one looks more carefully, it is possible to uncover details that point to the importance of demographic processes involving the north/south (N/S) axis of Europe. For instance, when conducting a principal component (PC) analysis (a statistical method of summarizing highly multidimensional data into the most important elements of variation), one output is the relative amount of variation explained by the individual PC axes (dimensions). In our analysis, PC1 explained 0.30% of the variation, whereas PC2 explained 0.15%. Both of these numbers are small by typical standards in population genetics and reflect the fact that there is very little variation among European subgroups (recall humans are all very genetically similar; see Chakravarti 2014). However, the percentage

for PC1 is twice that for PC2, arguing that the north northwest/south southeast axis explains more of the genetic diversity. However, these results must be interpreted with great caution. PC axis directions are very difficult to interpret, in part, because they can be very sensitive to how uniformly individuals are sampled and analyzed with regard to genetic ancestry (Novembre and Stephens 2008; McVean 2009). Indeed, a study that shortly followed ours found PC1 ran west/east, largely because they had larger numbers of individuals sampled along an east–west axis (Heath et al. 2008). A more robust indicator of a roughly N/S axis are gradients in haplotype diversity (haplotypes are particular combinations of sequence variants at multiple locations along the sequence and tend to have greater resolution with regard to the inference of demographic processes compared with variants at any single site). In a second article on the POPRES sample (Auton et al. 2009), we described a N/S gradient in haplotype diversity and a similar result was reported by Lao et al. (2008).

Such a gradient in haplotype diversity could be consistent with (a) serial founder effects in an initial northward expansion from the south of Europe during the Paleolithic, (b) subsequent expansions from the south after the retreat of the glaciers at the end of the LGM, (c) a northward expansion of agriculturalists in the early Neolithic, (d) admixture across the Mediterranean from North Africa, or (e) simply higher sustained population sizes through time in the south attributable to better long-term survival conditions. In our article, we argued that admixture across the Mediterranean certainly contributes to the N/S pattern (process d), and more extensive work by others supports this hypothesis (Botigué et al. 2013). To what extent this admixture is layered on top of other patterns (caused by processes a, b, c, and e listed above) is still a very open question and shows the challenges of interpreting the results of a palimpsest pattern of genetic variation.

Although, generally, the level of genetic similarity among European individuals tends to be highly correlated with their geographic origin, there are certain groups of individuals, termed "population isolates," such as Sardinians, who because of particular features (e.g., a distinct cultural trait or topographical barrier) may be substantially more isolated from their neighbors than average. One of the exciting aspects of genome-wide data has been the ability to analyze these isolates in more detail and tease out patterns of genetic isolation not clearly visible using smaller amounts of genetic data. In one study, we aimed to genetically characterize one of the less-studied population isolates in Europe, the Sorbs of eastern Germany (Veeramah et al. 2011). The Sorbs are of interest to historians as they speak a West Slavic language and have maintained their traditional Slavic customs but have found themselves surrounded by Germanic speakers because of the complex population movements that occurred during the migration period of Europe. To provide a control, we compared the Sorbs to two relatively well-studied population isolates, the French Basques and Sardinians. Basque country is a mountainous region along the France/Spain border that lies in the epicenter of one the major refugia of the LGM (Achilli et al. 2004). Approximately 25% of people from the region speak a language isolate, Euskara, which has been proposed to descend from the languages present in Western Europe before the arrival of Indo-European speakers. As such, it has been hypothesized that the Basque people may have remnants of the genetic diversity of the upper Paleolithic era (Cavalli-Sforza 1988). Similar hypotheses have been advanced for the people on Sardinia (Cavalli-Sforza et al. 1994), where the Mediterranean Sea has presented a formidable barrier to large-scale immigration.

Consistent with their medieval origins, the Sorbs showed greater genetic affinity to individuals from other western Slavic–speaking populations, such as Poles and Czechs, than they did to neighboring Germans. However, the evidence for genetic isolation in the Sorbs was very subtle compared with that observed for the Basques and Sardinians, who were clear outliers from the general European pattern of genetic variation and consistent with the presence of much stronger and older barriers to gene flow. The finding of Basque genetic isolation was particularly interesting given that two previous studies (Garagnani et al. 2009; Laayouni et al. 2010) could not find any evidence of differentiation of Basque populations from their Spanish and French neighbors using a much smaller number of genetic markers (approximately 100 compared with the approximately 30,000 we used). Thus, this study showed the considerable power of genomic data for studying human European populations.

CLASSICAL STUDIES OF EUROPEAN GENETIC DIVERSITY USING UNIPARENTAL MARKERS

Genome-wide array data have clearly proved to be a very powerful tool for studying European genetic diversity. However, building on earlier studies that examined classical markers, such as blood groups and protein polymorphisms (Cavalli-Sforza et al. 1994), most of our understanding of the peopling of Europe based on genetic studies from the last approximately 25 years comes from studying the non-recombining portion of the Y chromosome (NRY) and mitochondrial DNA (mtDNA). These markers are simple to study because they are transmitted exclusively down the paternal and maternal lines, respectively (they are "uniparental" markers).

A common framework for examining uniparental data is to assign individuals to "haplogroups" (a reference haplotype) based on whether they share a common set of genetic variants along the known NRY or mtDNA tree. The distribution of mtDNA haplogroups across Europe has been shown to be remarkably uniform with the presence of three major clades: (H,V), (J,T), and (U3,U4,U5,K) (Torroni et al. 2006; Underhill and Kivisild 2007). Haplogroup H is the most abundant across Europe with frequencies ranging from 40% to 60%. Interestingly, its frequency is greatest in Basques, and it is believed that subclades such as H1 and H3 all originated from this single region and expanded out across Europe at the end of the LGM (Achilli et al. 2004). The NRY appears to be comparatively much more geographically structured with certain haplogroups having a peak frequency in a particular geographical region, perhaps reflecting the greater level of patrilocality touched on earlier (Seielstad et al. 1998). For example, the Y-chromosome haplogroup R1b has frequencies that are very high in Western Europe, whereas haplogroup R1a tends to be more common in the east. Because of their apparent clinal distribution from east to west, haplogroups J, E1b1b (Semino et al. 2004), and R1b (Balaresque et al. 2010) have all been suggested to have once been carried by male Neolithic farmers, although such theories are not necessarily universally accepted (e.g., see Busby et al. 2012).

The NRY and mtDNA have also proved useful for examining much more specific migration events from the historical (rather than prehistorical) era (Jobling 2012). For example, Zalloua and colleagues (2008) argue that the Phoe-nicians, a culture that exerted great influence on the Mediterranean during the 1st century BCE (Before the Common Era), contributed >6% of paternal lineages to modern-day populations with a history of Phoenician contact.

ANCIENT DNA STUDIES IN EUROPE

Although there are considerable challenges with regard to degradation and contamination of DNA samples, the study of ancient DNA (paleogenetics) (Pääbo et al. 2004) is a promising avenue for understanding the peopling of Europe because it provides a direct account of diversity *in the past*. The vast majority of studies to date have exclusively examined mtDNA because, as thousands of copies are present per cell as compared with just one copy for nuclear DNA, it tends to be much more abundant in ancient remains. Generally, it appears there is substantial discontinuity between both Paleolithic and Neolithic as well as (though to a lesser extent) Neolithic and modern mtDNA pools in central and northern Europe (Bramanti et al. 2009; Malmström et al. 2009; Haak et al. 2010). There has, therefore, likely been significant post-Neolithic reshaping of European maternal lineages (e.g., during the metal ages) (Deguilloux et al. 2012), although greater mtDNA Neolithic continuity has been observed in Western Europe (Sampietro et al. 2007; Lacan et al. 2011).

However, mtDNA offers a view of genetic diversity that is limited to maternal lineages only. Fortunately, the development of next-generation sequencing has been revolutionary for paleogenetics (Stoneking and Krause 2011). Whole genomes from a Neanderthal (Green et al. 2010) and other archaic humans (Reich et al. 2010), as well as an approximately 4000-year-old Paleo-Eskimo from Greenland (Rasmussen et al. 2010), have now been sequenced; very recently, we saw the publication of the complete genome of a human European mummy commonly referred to as the Tyrolean Iceman or "Ötzi" (Keller et al. 2012).

Ötzi was discovered in the Eastern Alps near the Austrio-Italian border and dates back to the Late Neolithic/Early Copper Age (5350–5100 thousand years ago). Interestingly, Ötzi's genome clustered with Sardinians, some 500 miles away from where the sample was found. Thus, the study points to a significant change in the geographic distribution of genetic diversity since the Late Neolithic, and that modern Sardinians have likely conserved some of this ancient heritage. Two other research groups have since managed to sequence more than 10 million base pairs from late hunter–gatherers and Neolithic farmers in Scandinavia (Skoglund et al. 2012), as well as two Mesolithic hunter–gatherers in northwestern Spain (Sánchez-Quinto et al. 2012). It seems likely that other such studies will emerge soon giving us an unparalleled view of the processes that have led to the current distribution of European genetic diversity.

DARWINIAN SELECTION WITHIN EUROPE

Thus far, we have discussed the major patterns of variation that we observe when looking at genetic regions that are mostly presumed to be unaffected by positive natural selection. As mentioned above, the vast majority of human genetic diversity is thought to be shaped primarily by demographic processes such as population growth, dispersal patterns, and migrations. Some very small fraction of variants are, however, new mutations that rapidly increase in frequency because they confer some reproductive advantage to individuals that carry them resulting in their higher evolutionary fitness. One hypothesis for their scarcity in our genome is that the probability of a mutation being beneficial for an organism, rather than being deleterious or selectively neutral, is very small. However, the scale of population genetic surveys over the past decade has made it possible for researchers to identify some of these selectively adaptive variants among the much larger sea of variants that are mainly impacted by demographic history.

The selected variants discovered thus far seem to fall into three major biological categories: variants affecting human immunity, human external morphology (such as hair and eye color), and human dietary metabolism.

For example, one of the most dramatic variants to have been discovered in Europeans occurs in a locus affecting lactase persistence, *LCT*. Using metrics that describe allele frequency differences and/or reduced haplotype diversity, the pattern of genetic diversity at this gene sticks out among the backdrop of European diversity like a skyscraper from a prairie (Fig. 3). In this region, there is a single haplotype that is very common in many populations (Bersaglieri et al. 2004; Itan et al. 2010), presumably because it carries an advantageous variant. Specifically, it is thought the underlying functional mutation confers to its carriers the considerable nutritional benefit of the ability to digest milk more easily as adults.

A handful of genes affecting external morphological traits, such as eye color and skin pigmentation, also show similarly extreme patterns. For example, *SLC45A2* (Lucotte and Yuasa 2011) affects skin pigmentation and possesses a genetic variant that is found at very different frequencies across the globe. In these cases, novel variants have been driven to high frequency because they facilitate the interaction of their carriers with their physical environment (Jablonski 2008).

There have also been many pathogens that have ravaged Europe across time, and variants at genes underlying adaptive immunity, such as the *MHC* (Meyer and Thomson 2001), and innate immunity, such as *TLR6* (Pickrell et al. 2009), show signatures of positive selection, which are likely the result of adaptation to these unique pathogen exposures. A particular gene of interest is *CCR5*, which harbors a 32-base pair deletion known to confer resistance to HIV in humans, and in which the highest local frequency is found in Europe (16%). Some have suggested that its unique frequency in Europe is because the variant also conferred resistance to the bubonic plague that ravaged Europe during the Middle Ages, but the story appears to be much more complex and may not involve resistance to an ancestral pathogen at all (see review in Novembre and Han 2012).

The selection pressures just described are not unique to Europe. For example, in East Africa, other instances of the lactase persistence

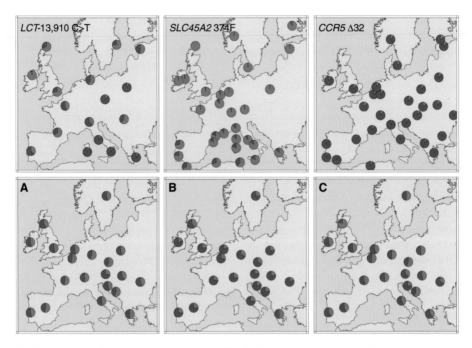

Figure 3. Maps showing the frequency distribution of individual genetic variants in European populations. The *top* row shows three variants in the genes *LCT*, *SLC45A2*, and *CCR5* that are thought to be under positive selection (selected variants are the green wedges). The *bottom* row shows three randomly chosen variants (*A–C*) with a minor allele frequency of >5% in all Europeans. It is noteworthy that the *top* row shows a high level of regional structuring (i.e., the selected allele for *LCT* is generally at much higher frequency in the northwest), whereas variants in the *bottom* row show a relatively flat distribution across the continent. The *LCT* data is from Itan et al. (2010), *SLC45A2* data is from Lucotte and Yuasa (2011), and *CCR5* data is from Novembre et al. (2005). In addition, all selected variants are supplemented by data from the ALFRED database (Kidd et al. 2003). The three random variants are based on frequencies in the POPRES sample.

variant have arisen in the *LCT* genomic region, and, in high latitudes of Asia, other variants in different genes that lighten skin pigmentation have spread to high frequency (see review in Gomez et al. 2014). Therefore, it seems that when distant human populations are faced with similar selection pressures, independent evolutionary responses can take place.

CONCLUDING REMARKS

Looking to the future, we expect that studies of European genetic diversity will continue to untangle how people settled and migrated through Europe and adapted to their environment. Although this could all be considered fascinating simply from a purely historical or anthropological perspective, there are also important cur-

rent practical applications of this knowledge. As previously mentioned, the development and widespread use of high-throughput array-based genotyping was largely driven by research comparing sets of disease cases with sets of control individuals at variants from across the genome. Differences in genetic ancestry, even minor ones (Novembre et al. 2008), between cases and controls can lead to significant numbers of spurious signals of apparent disease association, especially when examining many genetic markers (Hirschhorn and Daly 2005). One of the first examples of this for Europeans was a study that showed highly significant association between height and the variant most commonly associated with lactase persistence in Europeans, *LCT* C-13910T. However, this result was simply an artifact of population genetic stratification because Europeans of both northwest and southeast genetic ancestry were present in the cohort, and accounting for this feature resolved this artifact (Campbell et al. 2005). Thus, a good understanding of the patterns of genetic diversity observed in present-day Europeans (as well as others of recent European ancestry) is invaluable, both for statistically correcting for differences in ancestry and the initial choice of study population(s).

Understanding European genetic diversity is also important for predicting the ancestry of genetic data from samples of unknown European origin. Clearly accurate inference is of great value for forensic applications, but there is also considerable interest from the public, from a more personal or recreational viewpoint, with the geographic ancestry testing services of companies such as Family Tree DNA, 23andMe, and Ancestry.com. Although the issues involved are too complex to discuss here (e.g., Royal et al. 2010), in light of the unregulated nature of this industry, great care needs to be taken in the interpretation of these ancestry results. This is especially true today as many companies have shifted from examining NRY and mtDNA lineages only, which are limited in power and scope but yield relatively simple genealogical interpretations, to ancestry estimation from whole genome data. Although powerful insights can be gained regarding populations, inferences drawn from the analysis of any one particular individual's data can be highly misleading; for example, an individual interpreted as being of central European ancestry can, in fact, be of mixed northern and southern European ancestry. However, if applied appropriately, there are potentially exciting opportunities that can empower individuals who desire more information about their ancestors beyond that which is available from more traditional historical sources.

ACKNOWLEDGMENTS

Support for this work was provided by the National Institutes of Health to K.R.V. (R01-HG005226) and J.N. (R01-HG007089), as well as by the Searle Scholar Program (J.N.) and the National Science Foundation (DBI-0933731 to J.N.).

REFERENCES

*Reference is also in this collection.

Achilli A, Rengo C, Magri C, Battaglia V, Olivieri A, Scozzari R, Cruciani F, Zeviani M, Briem E, Carelli V, et al. 2004. The molecular dissection of mtDNA Haplogroup H confirms that the Franco-Cantabrian glacial refuge was a major source for the European gene pool. Am J Hum Genet 75: 910–918.

Auton A, Bryc K, Boyko AR, Lohmueller KE, Novembre J, Reynolds A, Indap A, Wright MH, Degenhardt JD, Gutenkunst RN, et al. 2009. Global distribution of genomic diversity underscores rich complex history of continental human populations. Genome Res 19: 795–803.

Balaresque P, Bowden GR, Adams SM, Leung HY, King TE, Rosser ZH, Goodwin J, Moisan JP, Richard C, Millward A, et al. 2010. A predominantly Neolithic origin for European paternal lineages. PLoS Biol 8: e1000285.

Barbujani G, Goldstein DB. 2004. Africans and Asians abroad: Genetic diversity in Europe. Annu Rev Genomics Hum Genet 5: 119–150.

Bersaglieri T, Sabeti PC, Patterson N, Vanderploeg T, Schaffner SF, Drake JA, Rhodes M, Reich DE, Hirschhorn JN. 2004. Genetic signatures of strong recent positive selection at the lactase gene. Am J Hum Genet 74: 1111–1120.

Botigué LR, Henn BM, Gravel S, Maples BK, Gignoux CR, Corona E, Atzmon G, Burns E, Ostrer H, Flores C, et al. 2013. Gene flow from North Africa contributes to differential human genetic diversity in southern Europe. Proc Natl Acad Sci 110: 11791–11796.

Bramanti B, Thomas MG, Haak W, Unterlaender M, Jores P, Tambets K, Antanaitis-Jacobs I, Haidle MN, Jankauskas R, Kind CJ, et al. 2009. Genetic discontinuity between local hunter–gatherers and central Europe's first farmers. Science 326: 137–140.

Busby GB, Brisighelli F, Sánchez-Diz P, Ramos-Luis E, Martinez-Cadenas C, Thomas MG, Bradley DG, Gusmão L, Winney B, Bodmer W, et al. 2012. The peopling of Europe and the cautionary tale of Y chromosome lineage R-M269. Proc Biol Sci 279: 884–892.

Campbell CD, Ogburn EL, Lunetta KL, Lyon HN, Freedman ML, Groop LC, Altshuler D, Ardlie KG, Hirschhorn JN. 2005. Demonstrating stratification in a European American population. Nat Genet 37: 868–872.

Cavalli-Sforza LL. 1988. The Basque population and ancient migrations in Europe. Munibe 6: 129–137.

Cavalli-Sforza LL, Menozzi P, Piazza A. 1994. The history and geography of human genes. Princeton University Press, Princeton, NJ.

* Chakravarti A. 2014. Perspectives on human variation through the lens of diversity and race. Cold Spring Harb Perspect Biol doi: 10.1101/cshperspect.a023358.

Deguilloux MF, Leahy R, Pemonge MH, Rottier S. 2012. European neolithization and ancient DNA: An assessment. Evol Anthropol 21: 24–37.

Garagnani P, Laayouni H, González-Neira A, Sikora M, Luiselli D, Bertranpetit J, Calafell F. 2009. Isolated populations as treasure troves in genetic epidemiology: The case of the Basques. *Eur J Hum Genet* **17:** 1490–1494.

Geary PJ. 2003. *The myth of nations.* Princeton University Press, Princeton, NJ.

* Gomez F, Hirbo J, Tishkoff SA. 2014. Genetic variation and adaptation in Africa: Implications for human evolution and disease. *Cold Spring Harb Perspect Biol* **6:** a008524.

Green RE, Krause J, Briggs AW, Maricic T, Stenzel U, Kircher M, Patterson N, Li H, Zhai W, Fritz MH, et al. 2010. A draft sequence of the Neandertal genome. *Science* **328:** 710–722.

Haak W, Balanovsky O, Sanchez JJ, Koshel S, Zaporozhchenko V, Adler CJ, Der Sarkissian CS, Brandt G, Schwarz C, Nicklisch N, et al. 2010. Ancient DNA from European early neolithic farmers reveals their near eastern affinities. *PLoS Biol* **8:** e1000536.

Heath SC, Gut IG, Brennan P, McKay JD, Bencko V, Fabianova E, Foretova L, Georges M, Janout V, Kabesch M, et al. 2008. Investigation of the fine structure of European populations with applications to disease association studies. *Eur J Hum Genet* **16:** 1413–1429.

Hewitt GM. 1999. Post-glacial re-colonization of European biota. *Biol J Linn Soc* **68:** 87–112.

Hirschhorn JN, Daly MJ. 2005. Genome-wide association studies for common diseases and complex traits. *Nat Rev Genet* **6:** 95–108.

Itan Y, Jones BL, Ingram CJ, Swallow DM, Thomas MG. 2010. A worldwide correlation of lactase persistence phenotype and genotypes. *BMC Evol Biol* **10:** 36.

Jablonski NG. 2008. *Skin: A natural history.* University of California Press, Berkeley, CA.

Jobling MA. 2012. The impact of recent events on human genetic diversity. *Philos Trans R Soc Lond B Biol Sci* **367:** 793–799.

Jobling MA, Hurles M, Tyler-Smith C. 2003. *Human evolutionary genetics.* Garland Science, New York.

Keller A, Graefen A, Ball M, Matzas M, Boisguerin V, Maixner F, Leidinger P, Backes C, Khairat R, Forster M, et al. 2012. New insights into the Tyrolean Iceman's origin and phenotype as inferred by whole-genome sequencing. *Nat Commun* **3:** 698.

Kidd KK, Rajeevan H, Osier MV, Cheung KH, Deng H, Druskin L, Heinzen R, Kidd JR, Stein S, Pakstis AJ, et al. 2003. ALFRED the ALlele FREquency Database Update. *Am J Phys Anthropol* **36:** S128.

Laayouni H, Calafell F, Bertranpetit J. 2010. A genome-wide survey does not show the genetic distinctiveness of Basques. *Hum Genet* **127:** 455–458.

Lacan M, Keyser C, Ricaut FX, Brucato N, Duranthon F, Guilaine J, Crubézy E, Ludes B. 2011. Ancient DNA reveals male diffusion through the Neolithic Mediterranean route. *Proc Natl Acad Sci* **108:** 9788–9791.

Lao O, Lu TT, Nothnagel M, Junge O, Freitag-Wolf S, Caliebe A, Balascakova M, Bertranpetit J, Bindoff LA, Comas D, et al. 2008. Correlation between genetic and geographic structure in Europe. *Curr Biol* **18:** 1241–1248.

Lewontin RC. 1972. The apportionment of human diversity. *Evol Biol* **6:** 381–398.

Lucotte G, Yuasa I. 2011. Decreasing values, from the North of West Europe to North Africa, of 374F allele frequencies in the skin pigmentation gene *SLC45A2:* An analysis. *J Evol Biol Res* **3:** 29–32.

Malmström H, Gilbert MT, Thomas MG, Brandström M, Storå J, Molnar P, Andersen PK, Bendixen C, Holmlund G, Götherström A, et al. 2009. Ancient DNA reveals lack of continuity between neolithic hunter–gatherers and contemporary Scandinavians. *Curr Biol* **19:** 1758–1762.

McVean G. 2009. A genealogical interpretation of principal components analysis. *PLoS Genet* **5:** e1000686.

Meyer D, Thomson G. 2001. How selection shapes variation of the human major histocompatibility complex: A review. *Ann Hum Genet* **65:** 1–26.

Nelson MR, Bryc K, King KS, Indap A, Boyko AR, Novembre J, Briley LP, Maruyama Y, Waterworth DM, Waeber G, et al. 2008. The population reference sample, POPRES: A resource for population, disease, and pharmacological genetics research. *Am J Hum Genet* **83:** 347–358.

Novembre J, Han E. 2012. Human population structure and the adaptive response to pathogen-induced selection pressures. *Philos Trans R Soc Lond B Biol Sci* **367:** 878–886.

Novembre J, Stephens M. 2008. Interpreting principal component analyses of spatial population genetic variation. *Nat Genet* **40:** 646–649.

Novembre J, Galvani AP, Slatkin M. 2005. The geographic spread of the CCR5 Δ32 HIV-resistance allele. *PLoS Biol* **3:** e339.

Novembre J, Johnson T, Bryc K, Kutalik Z, Boyko AR, Auton A, Indap A, King KS, Bergmann S, Nelson MR, et al. 2008. Genes mirror geography within Europe. *Nature* **456:** 98–101.

Pääbo S, Poinar H, Serre D, Jaenicke-Despres V, Hebler J, Rohland N, Kuch M, Krause J, Vigilant L, Hofreiter M. 2004. Genetic analyses from ancient DNA. *Annu Rev Genet* **38:** 645–679.

Pickrell JK, Coop G, Novembre J, Kudaravalli S, Li JZ, Absher D, Srinivasan BS, Barsh GS, Myers RM, Feldman MW, et al. 2009. Signals of recent positive selection in a worldwide sample of human populations. *Genome Res* **19:** 826–837.

Ralph P, Coop G. 2013. The geography of recent genetic ancestry across Europe. *PLoS Biol* **11:** e1001555.

Rasmussen M, Li Y, Lindgreen S, Pedersen JS, Albrechtsen A, Moltke I, Metspalu M, Metspalu E, Kivisild T, Gupta R, et al. 2010. Ancient human genome sequence of an extinct Palaeo-Eskimo. *Nature* **463:** 757–762.

Reich D, Green RE, Kircher M, Krause J, Patterson N, Durand EY, Viola B, Briggs AW, Stenzel U, Johnson PL, et al. 2010. Genetic history of an archaic hominin group from Denisova Cave in Siberia. *Nature* **468:** 1053–1060.

Renfrew C. 2010. Archaeogenetics—Towards a "new synthesis"? *Curr Biol* **20:** R162–R165.

Royal CD, Novembre J, Fullerton SM, Goldstein DB, Long JC, Bamshad MJ, Clark AG. 2010. Inferring genetic ancestry: Opportunities, challenges, and implications. *Am J Hum Genet* **86:** 661–673.

Sampietro ML, Lao O, Caramelli D, Lari M, Pou R, Martí M, Bertranpetit J, Lalueza-Fox C. 2007. Palaeogenetic evidence supports a dual model of Neolithic spreading into Europe. *Proc Biol Sci* **274:** 2161–2167.

Sánchez-Quinto F, Schroeder H, Ramirez O, Avila-Arcos MC, Pybus M, Olalde I, Velazquez AM, Marcos ME, Encinas JM, Bertranpetit J, et al. 2012. Genomic affinities of two 7,000-year-old Iberian hunter–gatherers. *Curr Biol* **22:** 1494–1499.

Seielstad MT, Minch E, Cavalli-Sforza LL. 1998. Genetic evidence for a higher female migration rate in humans. *Nat Genet* **20:** 278–280.

Semino O, Magri C, Benuzzi G, Lin AA, Al-Zahery N, Battaglia V, Maccioni L, Triantaphyllidis C, Shen P, Oefner PJ, et al. 2004. Origin, diffusion, and differentiation of Y-chromosome haplogroups E and J: Inferences on the neolithization of Europe and later migratory events in the Mediterranean area. *Am J Hum Genet* **74:** 1023–1034.

Skoglund P, Malmström H, Raghavan M, Storå J, Hall P, Willerslev E, Gilbert MT, Götherström A, Jakobsson M. 2012. Origins and genetic legacy of Neolithic farmers and hunter–gatherers in Europe. *Science* **336:** 466–469.

Stoneking M, Krause J. 2011. Learning about human population history from ancient and modern genomes. *Nat Rev Genet* **12:** 603–614.

Torroni A, Achilli A, Macaulay V, Richards M, Bandelt HJ. 2006. Harvesting the fruit of the human mtDNA tree. *Trends Genet* **22:** 339–345.

Underhill PA, Kivisild T. 2007. Use of Y chromosome and mitochondrial DNA population structure in tracing human migrations. *Annu Rev Genet* **41:** 539–564.

Veeramah KR, Tönjes A, Kovacs P, Gross A, Wegmann D, Geary P, Gasperikova D, Klimes I, Scholz M, Novembre J, et al. 2011. Genetic variation in the Sorbs of eastern Germany in the context of broader European genetic diversity. *Eur J Hum Genet* **19:** 995–1001.

Wall JD, Yang MA, Jay F, Kim SK, Durand EY, Stevison LS, Gignoux C, Woerner A, Hammer MF, Slatkin M. 2013. Higher levels of Neanderthal ancestry in East Asians than in Europeans. *Genetics* **194:** 199–209.

Zalloua PA, Platt DE, El Sibai M, Khalife J, Makhoul N, Haver M, Xue Y, Izaabel H, Bosch E, Adams SM, et al. 2008. Identifying genetic traces of historical expansions: Phoenician footprints in the Mediterranean. *Am J Hum Genet* **83:** 633–642.

Genetic Variation and Adaptation in Africa: Implications for Human Evolution and Disease

Felicia Gomez[1,2,3,4], Jibril Hirbo[1,3,5], and Sarah A. Tishkoff[1]

[1]Department of Genetics and Biology, School of Medicine and School of Arts and Sciences, University of Pennsylvania, Philadelphia, Pennsylvania 19104

[2]Hominid Paleobiology Doctoral Program and The Center for the Advanced Study of Hominid Paleobiology, Department of Anthropology, The George Washington University, Washington, D.C. 20052

Correspondence: tishkoff@mail.med.upenn.edu

Because modern humans originated in Africa and have adapted to diverse environments, African populations have high levels of genetic and phenotypic diversity. Thus, genomic studies of diverse African ethnic groups are essential for understanding human evolutionary history and how this leads to differential disease risk in all humans. Comparative studies of genetic diversity within and between African ethnic groups creates an opportunity to reconstruct some of the earliest events in human population history and are useful for identifying patterns of genetic variation that have been influenced by recent natural selection. Here we describe what is currently known about genetic variation and evolutionary history of diverse African ethnic groups. We also describe examples of recent natural selection in African genomes and how these data are informative for understanding the frequency of many genetic traits, including those that cause disease susceptibility in African populations and populations of recent African descent.

Africa is where modern humans evolved and is the starting place for the global expansion of our species (Stringer and Andrews 1988; Stringer 1994; Templeton 2002). African populations also have the highest levels of genetic and phenotypic variation among all humans. This variation is informative for characterizing demographic history in Africa, including times when populations increased in size, contracted, migrated, or when admixture between them occurred. Genetic variation also provides data that are useful in identifying local adaptation to diverse environments (Campbell and Tishkoff 2008). Characterizing human genetic variation and examining phenotypic variation in extant African populations is fundamental to the identification of genes that play a role in function, adaptation, and complex disease susceptibility

---

[3]These authors contributed equally to this work.

[4]Current address: Division of Statistical Genetics, Department of Genetics, Washington University School of Medicine in St. Louis, St. Louis, Missouri 63108.

[5]Current address: Department of Biological Sciences, Vanderbilt University, 7280 BSB/MRB III, VU Station B, Box 35-1634, Nashville, Tennessee 37235-1634.

Cite this article as *Cold Spring Harb Perspect Biol* doi: 10.1101/cshperspect.a008524

in Africans and populations of recent African descent.

LINGUISTIC, CULTURAL, AND SUBSISTENCE DIVERSITY IN AFRICA

There are over 2000 indigenous languages spoken in Africa. They are classified into four major linguistic families—Niger-Kordofanian (spoken primarily by agriculturalist populations located in large contiguous regions of sub-Saharan Africa from West Africa to eastern and southern Africa), Nilo-Saharan (spoken predominantly by pastoralist populations in central and eastern Africa), Afroasiatic (spoken predominantly by agro-pastoralists and pastoralist populations in northern and eastern Africa), and Khoesan (a language family that contains click consonants, spoken by hunter–gatherer San populations in southern Africa as well as the Hadza and Sandawe hunter–gatherers in Tanzania) (Fig. 1).

African populations also live in a diverse range of environments (including deserts, savannahs, and tropical environments), have different exposures to infectious disease, and have diverse diets and subsistence patterns. Because African populations exist in such a broad spectrum of environments, they may have regional or population-specific genetic variants that play a role in local adaptation to different selection pressures. The study of African genomic variation is uniquely interesting because of the large proportion of human genetic variation observed in African populations and the length of human history in the continent.

PATTERNS OF GENETIC DIVERSITY AND DEMOGRAPHIC HISTORY IN AFRICA

The characterization of human genetic variation began with studies that described human blood groups and protein polymorphisms (Cavalli-Sforza 1994). Since then, it has grown to include numerous studies of mtDNA (mitochondrial DNA) that is inherited maternally, Y chromosome variation that is inherited paternally, and loci from the full nuclear genome that include sequences of repetitive DNA, single nucleotide polymorphisms (SNPs), and struc-

tural variation (SVs) (Campbell and Tishkoff 2008).

The pattern of genetic variation at these loci in modern African populations reflects their demographic and evolutionary histories (Campbell and Tishkoff 2008). Modern humans evolved in Africa ∼200 kya (thousand years ago) and migrated from Africa within the past 50–100 kya, successfully colonizing most of the terrestrial parts of the globe (Campbell and Tishkoff 2008). This is called the "Out-of-Africa" migration. Currently, the precise location of the origin of modern humans remains a contentious issue. Some argue that South Africa is the location where our species originated (Tishkoff et al. 2009; Compton 2011; Henn et al. 2011), whereas others argue that East Africa is where modern humans originated (Prugnolle et al. 2005; Ray et al. 2005). However, these inferences are biased by the lack of archeological and fossil data in tropical areas of Africa and the fact that the geographic location of populations in the present may have differed in the past.

Genetic evidence also indicates that human populations underwent several major population expansions at different time periods over the last 100 ky (Rogers and Harpending 1992; Harpending et al. 1993; Excoffier and Schneider 1999; Atkinson et al. 2009; Cox et al. 2009): an initial major population expansion in Africa ∼110–70 kya (Rogers and Harpending 1992; Harpending et al. 1993; Excoffier and Schneider 1999; Atkinson et al. 2008, 2009) and subsequent migration and expansion events around the globe at ∼60–55 kya (Atkinson et al. 2009), ∼40–25 kya, and ∼12 kya (Harpending et al. 1993; Atkinson et al. 2009; Cox et al. 2009).

Prior to the domestication of plants and animals ∼10,000 years ago, all human populations practiced hunting-gathering/foraging for subsistence. Although we do not know their original homeland, ancestors of the extant African hunter–gatherer populations, including the current Khoesan speakers, migrated into central, eastern, and southern Africa probably >20 kya (Nurse 1997; Ehret 2002). It is speculated that speakers of Khoesan languages were widespread throughout a contiguous area that encompassed southern and eastern African re-

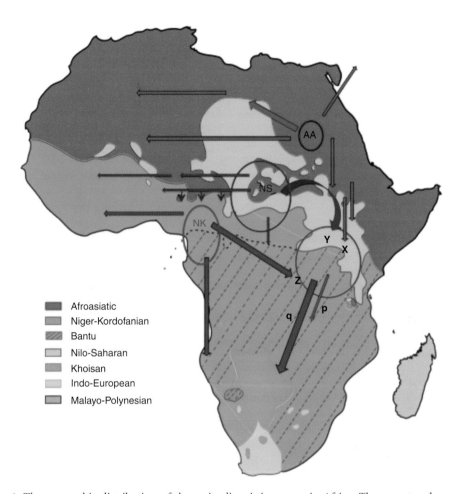

Figure 1. The geographic distribution of the major linguistic groups in Africa. The map was drawn using information (geographic locations of ethnic speakers in Africa that are based on published sources) from Haspelmath et al. (2008) (wals.info/), www.ethnologue.com (International HapMap 2005), Greenberg (1963, 1972), Vansina (1995), and Ehret (1971, 1993, 1995, 2001b). Geographic range occupied by Bantu speakers, the major linguistic subfamily within the Niger-Kordofanian phylum mentioned in the text, is also shown. Putative centers of origin and estimated time of initial expansion based on linguistic studies for three of the four language families are also listed: AA, Afroasiatic (14 kya) (Ehret 1995); NS, Nilo-Saharan (Blench 1993; Ehret 1993; Blench 2006); NK, Niger-Kordofanian (5 kya) (Nurse 1997; Ehret 2001a). Afroasiatic-speaking pastoralists were the first food-producing populations to migrate into East Africa circa 5 kya (X) (Leakey 1931; Butzer 1969; Robbins 1972; Barthelme 1977); followed by Nilo-Saharan-speaking pastoralists circa 3 kya (Y) (Leakey 1931; Bower 1973; Ambrose 1982; Distefano 1990), and later Bantu-speaking agriculturalists after circa 2.5 kya (Z) (Posnansky 1961a,b). P and q represent initial expansion of pastoralists (2.5 kya) and later Bantu-speaking agriculturalists (after 2 kya) to southern Africa from East Africa, respectively.

gions (Sutton 1973; Smith 1992). Based on archaeological and linguistic data, it is hypothesized that recent migration events that occurred, due to changes in subsistence patterns, modified the genetic landscape in Africa through admixture of populations practicing pastoralism and/ or agriculture with indigenous hunter–gatherer populations. For example, the expansion of agriculturalists from Central-West Africa and the migration of pastoralist populations from northeastern Africa have contributed to the current genetic landscape in Africa. One of the largest migration events in Africa that was mediated by a shift in subsistence patterns was the migration of Bantu-speaking populations, a subfamily of Niger-Kordofanian languages, from their proposed homeland in Nigeria/Cameroon across sub-Saharan Africa within the past 5000 years (Fig. 1) (Nurse 1997; Ehret 2001a). The result of this expansion is the current predominance of Bantu languages in many African countries. Other major migration events have included the migration of Nilo-Saharan pastoralists from their homeland in Chad/Sudan, both westward across the Sahel ~7000 years ago and eastward into Kenya and Tanzania within the past ~3000 years (Blench 1993; Ehret 1993). Afro-Asiatic-speaking agro-pastoralists migrated from their homeland, which encompassed the Nile Valley to the Ethiopian highlands, into western, northern, and eastern Africa within the past 8000–5000 years (Ehret 1995, 1998) (Fig. 1). Demographic history in Africa has also been influenced by the more recent migration of non-Africans into Africa. Most noteworthy is the migration of southwestern Asians into Northern and Eastern Africa within the past ~3000 years (Van Beek 1967; Butzer 1981; Phillipson 2009). Europeans, southern and eastern Asians have also migrated into Southern Africa within the past several hundred years.

Consistent with the "Out of Africa" model of modern human origins, analyses of genetic data indicate that Africans have higher levels of genetic diversity than non-Africans (Cann et al. 1987, 2002; Tishkoff et al. 1998; Marth et al. 2004). Overall, the extant patterns of genetic variation in global populations suggests serial founder events, with increasing genetic distance correlated with geographic distance from East Africa, indicating the cumulative effect of genetic drift as humans expanded into the rest of the world (Prugnolle et al. 2005; Ramachandran et al. 2005; Hofer et al. 2009). Studies of autosomal genetic variation show that genetic distance between populations also increases with geographic distance within Africa (Prugnolle et al. 2005; Ramachandran et al. 2005; Hofer et al. 2009; Tishkoff et al. 2009). The greater genetic diversity in sub-Saharan Africans, as compared to other continental populations, may also be partially a result of ancient admixture with yet-to-be identified archaic population(s) (Zietkiewicz et al. 1998; Labuda et al. 2000; Plagnol and Wall 2006; Hammer et al. 2011; Lachance et al. 2012), analogous to the proposed admixture between Neanderthals and modern humans in Eurasia (Reich et al. 2010, 2011; Abi-Rached et al. 2011; Green et al. 2011; Yotova et al. 2011). Genomic introgression from Neanderthals has been detected in some East African populations but was inferred to be a signature of recent back-migrations from Eurasia (Wang et al. 2013).

Many of the recent population genetic studies of African populations are based on analysis of genetic markers that were genotyped in a small number of populations that are part of the CEPH-HGDP collection (Centre d'Étude du Polymorphisme Humain–Human Genome Diversity Panel) or the International Haplotype Map (HapMap) Project (Batzer and Deininger 2002; Rosenberg et al. 2002; Li et al. 2008). Only six African populations are included in the CEPH-HGDP panel, whereas the International HapMap Project includes just three African populations (Niger-Kordofanian-speaking Yoruba from Nigeria, Bantu-speaking Luhya from Kenya, and Nilo-Saharan-speaking Maasai from Kenya). A more recent international collaborative effort that aims to catalogue human genetic variation through whole genome resequencing, the "The 1000 Genomes Project" (2010), originally had only two African populations represented in the project but has since expanded to include the Esan from Nigeria, individuals from The Gambia, the Luhya from Kenya, and the Mende from Sierra Leone. However, it should be noted that all of these populations speak Niger-Kordofanian languages and share recent genetic ancestry. The 1000 Genomes project has also expanded to include people of recent African descent including African Americans from the southwestern United States and African Caribbean individuals from Barbados. Although these projects are important in their description of overall human genetic diversity, they are limited in their coverage of African populations, despite the recent additions (Consortium 2010). Because these projects have sampled a small number of African populations it is likely

Cite this article as *Cold Spring Harb Perspect Biol* doi: 10.1101/cshperspect.a008524

that a substantial portion of human genetic variation will be missed. Thus, it is important to continue to add African populations that are underrepresented in human genomic studies, particularly those speaking languages belonging to the other three language families in Africa. Analysis of genomic variation across a broad range of African populations will be useful for elucidating fine-scale population structure and demographic patterns in African populations.

The current studies of global human genetic variation suggest that human populations may have been divided into distinct subpopulations in Africa prior to the migration of modern humans Out-of-Africa ~100 kya (Labuda et al. 2000; Satta and Takahata 2004; Adeyemo et al. 2005; Plagnol and Wall 2006; Bryc et al. 2009; Tishkoff et al. 2009; Wall et al. 2009; de Wit 2010; Patterson 2010; Sikora 2010; Henn et al. 2012; Pagani et al. 2012; Schlebusch et al. 2012). For example, we (Tishkoff et al. 2009) identified 14 ancestral population clusters in Africa with four predominant clusters that broadly represent populations from major African geographical regions and those that speak the four African language families mentioned above. The remaining 10 are mainly restricted to specific geographic regions, languages, or in some cases, individual populations (e.g., Hadza hunter–gatherers). Other studies of African genetic diversity (Adeyemo et al. 2005; Bryc et al. 2009; de Wit 2010; Patterson 2010; Sikora 2010; Henn et al. 2012; Pagani et al. 2012; Schlebusch et al. 2012) have largely recapitulated the regional clusters (Tishkoff et al. 2009) while also adding some fine-scale genetic structure between populations from different ethnicities or populations that speak different languages within the regions studied.

Studies of genetic variation in Africa suggest that even though high levels of mixed ancestry are observed in most African populations, the genetic variation observed in Africa is broadly correlated with geography, language classification (Adeyemo et al. 2005; Bryc et al. 2009; Tishkoff et al. 2009) and subsistence classifications (Tishkoff et al. 2009). For example, genetic variation among Nilo-Saharan and Afroasiatic-speaking populations from both Central and

East Africa (Tishkoff et al. 2009; Bryc et al. 2009) reflect the geographic region from which they originated, and generally shows a complex pattern of admixture between these populations and the Niger-Kordofanian speakers who migrated into the region more recently. Consistent with linguistic evidence regarding the origin of Nilo-Saharan languages in the Chad/Sudan border, the highest proportion of Nilo-Saharan ancestry is observed among southern Sudanese populations (Tishkoff et al. 2009). Additionally, North/Northwest African populations are genetically differentiated from sub-Saharan African populations, but have considerable shared ancestry with East African Afroasiatic speakers (Tishkoff et al. 2009; Henn et al. 2012) and southwest Asian populations (Rosenberg et al. 2002; Behar et al. 2008, 2010; Li et al. 2008; Kopelman et al. 2009; Tishkoff et al. 2009; Atzmon et al. 2010; Bray 2010; Hunter-Zinck 2010; Henn et al. 2012), likely reflecting historic migrations from those source populations into northern Africa. Finally, several studies of genetic variation in the "colored" population from South Africa (Tishkoff et al. 2009; de Wit et al. 2010; Patterson et al. 2010) are consistent with the documented history of this population, which indicates that they have South African Khoesan, Niger-Kordofanian, European, South Asian, and Indonesian ancestors (Tishkoff et al. 2009; de Wit et al. 2010; Patterson et al. 2010).

mtDNA AND Y CHROMOSOME VARIATION

A refined view of uniparental genetic diversity in Africa has been hampered by a lack of extensive sampling in Africa and a lack of consensus on the proper method to estimate mutation rates that are essential to make inferences about human demography. The time to the most recent common ancestor (TMRCA) for mtDNA lineages, based on large datasets, ranges from 121.5 to 221.5 kya, (Ingman et al. 2000; Behar et al. 2008), which is older than most estimates of the Y chromosome TMRCA, which are based on patchy sampling within Africa (60–140 kya) (Thomason 1988; Cruciani et al. 2011; Wei et al. 2013). The fact that there has been limited sampling in Africa is further exemplified by recent results that include more African samples and infer that the TMRCA of the human Y chromosome is similar (120–200 kya) (Mendez et al. 2013) or even older (237–581 kya) than the TMRCA of mtDNA (Ingman et al. 2000; Behar et al. 2008; Francalacci et al. 2013; Mendez et al. 2013; Poznik et al. 2013). Generally, there is a lack of precision in TMRCA estimates reported by different studies using the same marker systems (Francalacci et al. 2013; Mendez et al. 2013; Poznik et al. 2013). Nonetheless, the consensus based on numerous studies, is that human populations exhibit structure (e.g., genetic differences among populations from different regions) based on both mtDNA and Y chromosome variation (Wallace et al. 1999; Ingman et al. 2000; Maca-Meyer et al. 2001; Herrnstadt et al. 2002; Mishmar et al. 2003).

The distribution pattern of uniparental lineages in Africa suggests ancient geographical and cultural subdivision in African. African populations exhibit greater diversity in the Y chromosome than those of other continents (Hammer 2002). Based on the latest classification of Y chromosome haplogroups (Karafet et al. 2008), A, B, E, and J haplogroups are found in Africans, whereas C, D, and the lineages descending from haplogroup F are almost exclusively observed outside of sub-Saharan Africa (Fig. 2). The oldest lineages, A and B, with a TMRCA over ∼75 kya, are present only in Africans. The younger lineages, E and J haplogroups, with a TMRCA less than 75 kya, are present in both African and non-African populations (Fig. 2) (Seielstad et al. 1998; Scozzari et al. 1999). Haplogroup E represents the great majority of Y chromosome haplotypes across Africa (Cruciani et al. 2002). Furthermore, the sublineages within these haplogroups exhibit unique regional distribution patterns within Africa (Cruciani 2002, 2010, 2011; Hammer 2002; Luis et al. 2004; Wood et al. 2005; Batini 2007, 2011; Berniell-Lee et al. 2009), with some restricted to specific geographical regions and/or language families (Fig. 3), consistent with substructuring of diverse African populations.

Analogous to variation observed on the Y chromosome, mtDNA haplotypes also show geographic structuring. African populations are characterized by having the oldest haplogroup lineages (L0–L5), with TMRCA over ∼80 kya (Bandelt et al. 1995, 2001; Chen et al. 1995, 2000; Graven et al. 1995; Soodyall et al. 1996; Watson et al. 1996, 1997; Alves-Silva et al. 2000; Torroni et al. 2001; Salas et al. 2002), as well as the M1 haplogroup, an L3 sublineage. Although lineages M, N, and all of their descending lineages with inferred TMRCA less than 80 kya are found in most global populations, they are observed almost exclusively outside of sub-Saharan Africa. There is also a unique distribution pattern of mtDNA haplotypes across Africa with some lineages restricted to specific geographical regions and/or language families (Fig. 3) (Newman 1980; Vigilant 1990, 1991; Chen et al. 1995, 2000; Ehret 1995, 2006; Watson et al. 1996, 1997; Rando et al. 1998; Krings et al. 1999; Richards et al. 2000; Pereira et al. 2001; Salas et al. 2002; Thomas et al. 2002; Destro-Bisol et al. 2004; Kivisild et al. 2004; Beleza et al. 2005; Coia et al. 2005; Jackson et al. 2005; Cerny et al. 2006, 2008, 2009; Gonzalez et al. 2006; Batini et al. 2007; Gonder et al. 2007; Behar et al. 2008; Coudray et al. 2008; Quintana-Murci et al. 2008, 2010; Castri et al. 2009; Coelho et al. 2009; Poloni et al. 2009; Saunier et al. 2009; Stefflova et al. 2009; Veeramah et al. 2010). Additionally, there is also a predominance of some Y chromosome and mtDNA lineages in specific regions of Africa. There is also sharing of many of these lineages between regions (Watson et al. 1996, 1997; Pereira et al. 2001; Salas et al. 2002; Beleza et al. 2005; Behar et al. 2008; Quintana-Murci et al. 2008), suggesting that there have been both recent and ancient migration events between these regions over the last 20 ky (Ehret 1967).

Despite the large amount of ethnic and genetic diversity observed in African relative to non-African populations, there has been limited sampling in Africa. Additional sampling and characterization of genetic diversity, particularly from underrepresented central, southern and eastern African countries will be crucial to deciphering fine-scale genetic history of genetically, culturally, and linguistically diverse African populations and for reconstructing both ancient and recent human evolutionary history.

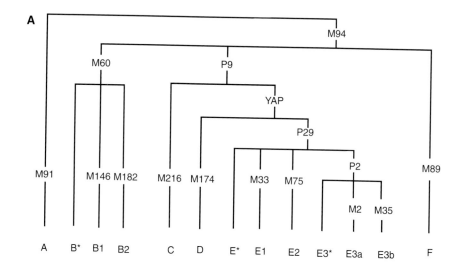

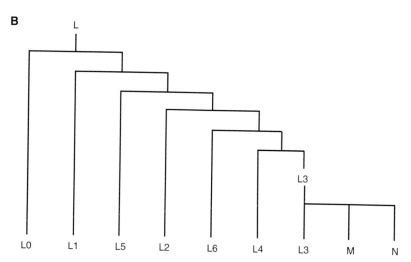

Figure 2. Evolutionary tree of Y chromosome and mtDNA haplogroups. (*A*) Nomenclature for major lineages of Y chromosome haplotypes. The M, P, and YAP labels leading to haplogroup/s are SNPs and indels that are used to define these haplogroups. Haplogroup F encompasses haplogroups F to T. Haplogroups A, B, and E are mainly found in Africa, whereas the rest are found mainly outside Africa. (*B*) Overview of mtDNA haplogroup phylogeny. In the mtDNA haplogroup nomenclature, the letter names of the haplogroups run from A to Z, with further subdivisions using numbers (from 0) and lowercase letters (from a). The naming was done in the order of their discovery and does not reflect the actual genetic relationships. Haplogroup M and N encompasses all of the haplogroups lettered A to Z excluding haplogroup L. Haplogroups L (L0–L6) are mainly found in Africa, whereas the rest are found mainly outside Africa (Richards et al. 1998; Macaulay et al. 1999; Quintana-Murci et al. 1999; Salas et al. 2002; Kivisild et al. 2004; Behar et al. 2008).

Comprehensive analysis of African genetic variation will help us address some of the outstanding questions about the origin of our species such as: (1) precisely where and when modern humans originated in Africa; (2) the number and age of ancestral structured populations in Africa; (3) the age and direction of ancient migration events both within and out of Africa (e.g., we do not yet have a consensus from genetic studies on the timing and route/s modern humans took out of Africa to populate the rest of the globe); and, (4) determination of the ar-

Figure 3. Distribution of mtDNA and Y chromosome haplogroups across Africa. The black lines represent the approximate geographical boundaries of the distribution of each haplogroup cluster, with each of the clusters predominantly observed in the demarcated regions.

chaic source population(s) from which genetic introgression occurred into African populations and when these events took place.

DARWINIAN SELECTION AND GENETIC VARIATION

Demographic events such as those described above, that is changes in population size, short- and long-range migrations, and admixture or gene flow from one population to another, influence the levels and patterns of genetic variation across the genome (Campbell and Tishkoff 2008). In addition to demography, natural selection, recombination and mutation, which take place in specific parts of the genome, can also influence genetic variability.

Natural selection occurs when the fitness effects of genotypes are unequal (Nielsen 2005). For example, a beneficial genetic variant is likely to increase in frequency in a population over time because of positive selection. Likewise, if a genetic variant is deleterious, it may be selected against and its frequency in a population will decrease over time. The occurrence of natural selection can thus create substantial differences in the patterns of genetic variation between human populations (Aquadro et al. 2001). Identifying loci that exhibit patterns of genetic variation that are indicative of natural selection is, therefore, important because these loci are likely to be regions of the genome that are, or have been, functionally important and play a role in adaptation to local environments. Furthermore, understanding which regions of the genome have been subject to natural selection will aid in the identification of genetic variation that contributes to phenotypic variation among human populations, including disease susceptibility (Ronald and Akey 2005). Here,

Cite this article as *Cold Spring Harb Perspect Biol* doi: 10.1101/cshperspect.a008524

we discuss results of several studies of local adaptation to distinct environments and diets within Africa.

Lactase Persistence

Adult mammals lose the ability to effectively digest lactose, the main carbohydrate in milk, after weaning. The inability to digest lactose is a result of decreased production of the enzyme lactase-phlorizin hydrolase, or lactase (Swallow 2003; Ingram et al. 2009). Individuals who are unable to digest lactose have what is called the "lactase nonpersistent" (LNP) trait. Although the LNP trait was once considered to be pathologic, it is now well understood that this trait is "ancestral" (i.e., all ancestors of modern humans had this condition) and widespread among human populations (Auricchio et al. 1963; Newcomer et al. 1983; Segal et al. 1983). Individuals with the "lactase persistent" (LP) trait continue to have high levels of lactase production into adulthood. The LP trait is common only in populations whose ancestors practiced cattle domestication or pastoralism (Ingram et al. 2009). For example, LP occurs at high frequencies in northern European dairying populations (e.g., Finns, Swedes, and Danes), and decreases in frequency in southern Europe, the Middle East, and in most of Asia (Durham 1991). LP is also found to be common in some pastoral populations in Africa (Tishkoff et al. 2007). Thus, it has been hypothesized that LP is an adaptive trait in human populations that practice cattle domestication and dairying (Swallow 2003).

In Europe, a common regulatory variant outside of the gene that encodes lactase (LCT) has been identified as being strongly associated with the LP trait (Enattah et al. 2002; Ingram et al. 2007; Tishkoff et al. 2007; Enattah et al. 2008). Several studies also suggest that this variant is causal and is a target of recent natural selection (Enattah et al. 2002; Poulter et al. 2003; Bersaglieri et al. 2004; Coelho et al. 2005). This suggestion is based on several lines of evidence. First, experiments in small intestine cell lines suggest that the European mutation increases the expression of the lactase gene and causes

an increase in the production of the lactase enzyme (Enattah et al. 2002; Olds and Sibley 2003; Troelsen et al. 2003; Lewinsky et al. 2005). Additionally, several different tests indicate that the LCT locus has experienced recent strong natural selection (Poulter et al. 2003; Bersaglieri et al. 2004; Coelho et al. 2005). These studies have identified unexpectedly long genomic regions of genetic similarity flanking the LCT locus, which is a signature of recent positive selection. When positive selection causes a variant or variants to increase in frequency, the selected variant will cause neighboring genetic variation to "hitchhike" as the selected variant increases in frequency, thereby causing large tracks of identical sequences surrounding the variant under selection. This genetic feature has been identified in many European individuals with the regulatory variant who have the LP trait (Poulter et al. 2003). Finally, the estimated age of the European LP-associated variant is between ~20,000 and ~2000 years ago (Bersaglieri et al. 2004; Coelho et al. 2005; Tishkoff et al. 2007). These dates are consistent with the origins of dairy farming in southwest Asia that have been estimated from the archaeological record at ~8–9 kya (Evershed et al. 2008).

Although this is a convincing example of gene/culture co-evolution in Europe, the genetic basis of the LP phenotype was largely unknown in Africa for many years. Previous studies (Mulcare et al. 2004) indicated that the European regulatory variant was noticeably absent in many African populations, including populations that have a history of pastoralism and drinking milk. This observation called into question whether the European variant is truly the causal variant for lactase persistence (Ingram et al. 2007). However, genotype/phenotype association studies of LP in Africa (Swallow 2003; Tishkoff et al. 2007) have identified three novel variants upstream of LCT, near the European variant, that are significantly associated with LP in African populations. All three of the African sites enhance the expression of LCT (Tishkoff et al. 2007; Enattah et al. 2008), and the most common of the three African mutations also shows strong evidence of recent strong positive selection. The estimated age of this African variant is ~3000–7000 years, consistent with the introduction of cattle domestication south of the Saharan Desert within the past ~5500 years (Tishkoff et al. 2007). Together, these examples from Europe and Africa show us that strong selective pressure can increase the frequency of numerous rare mutations that arise independently and regulate lactase gene expression where people practice dairying. The occurrence of multiple mutations that evolved under similar selection pressures is a good example of convergent evolution in several different human populations.

Malaria

Examination of population-level genetic variation and recent natural selection caused by infectious diseases can help us understand differences in disease susceptibility and incidence, and why some genetic variants, especially those that are potentially deleterious, are common in specific populations. Malaria is one such example that has had an important impact on patterns of human genetic variation and the geographic distribution of a number of genetic disorders (Fortin et al. 2002).

Malaria is a parasitic disease caused by species in the genus *Plasmodium*. There are several *Plasmodium* species that infect humans, including *P. falciparum, P. ovale, P. malariae,* and *P. vivax.* A large number of genetic variants that are associated with malaria-protective phenotypes have been identified and many of them cause serious disease (Ko et al. 2012; Gomez et al. 2013a). For example, variants associated with hemoglobinopathies, α and β thalassemias, and the structural hemoglobin variants S, C, and E, are classic examples of deleterious genetic variants that nevertheless confer protection from malaria (Weatherall 2001; Gomez et al. 2013b). Early studies of thalassemias (Haldane 1949; Flint et al. 1986) noted that these diseases have a geographic distribution that coincides with *P. falciparum* endemicity. In what is now called the "malaria hypothesis," JBS Haldane suggested that the genetic variants responsible for the thalassemias may be maintained in some human populations because of an advantage

associated with malaria resistance (Haldane 1949). In a recent analysis, Piel et al. (2010) revisited the global distribution of sickle cell anemia, a disease caused by a mutation in the β-globin gene (HbS). They showed strong geographical correlation between HbS allele frequency and malaria endemicity in Africa (Fig. 4), consistent with the malaria hypothesis.

Additional studies of other genes have shed further light on our understanding of the co-evolution of humans and *P. falciparum,* and provide evidence of recent natural selection in the human genome. For example, studies of genetic variation at *G6PD* (glucose-6-phosphate dehyrogenase), a gene that is known to harbor mutations associated with G6PD enzyme deficiency and protection from *P. falciparum* infection, have revealed signatures of genetic variation consistent with recent natural selection (Tishkoff et al. 2001; Saunders et al. 2002; Verrelli et al. 2002; Sabeti et al. 2006). Among the *G6PD* variants that are associated with protection from malaria, the A-variant is most common in sub-Saharan Africa. Several studies, in African and non-African populations (Tishkoff et al. 2001; Sabeti et al. 2002; Saunders et al. 2002; Verrelli et al. 2002), have identified signatures of recent natural selection acting on the A-variant and inferred the TMRCA for this variant to range from ~1200 to 12,000 ya, which is consistent with the idea that selective pressures caused by malaria increased in African populations as population densities increased after the introduction of domesticated plants and animals (Tishkoff et al. 2001).

Kidney Disease

In the United States, diseases like type 2 diabetes mellitus, hypertension, and prostate cancer are known to be more common in African Americans than European Americans (Landis et al. 1999; Brancati et al. 2000; Ong et al. 2007). This disparity in disease occurrence suggests that genetic factors mediating disease risk, together with environmental factors, may differ among European and African descent populations (Hunter 2005; Yang et al. 2005; Cooper 2013). It has been proposed that genetic risk

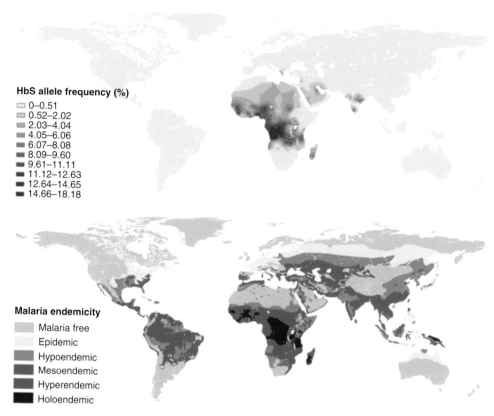

Figure 4. Global distribution of the HbS sickle cell anemia allele compared to the historic geographic distribution of malaria. (*Top*) Global distribution of HbS allele frequency predicted from Bayesian geostatistical modeling. (*Bottom*) Historical map of malaria endemicity. The classes are defined by $PfPr_{2-10}$ ($PfPr_{2-10}$ = proportion of 2- to 10-year olds with confirmed blood stage asexual parasites): malaria free, $PfPr_{2-10} = 0$; epidemic, $PfPr_{2-10} \approx 0$; hypoendemic, $PfPr_{2-10} < 0.10$; mesoendemic, $PfPr_{2-10} \geq 0.10$ and < 0.50; hyperendemic, $PfPr_{2-10} \geq 0.50$ and < 0.75; holoendemic, $PfPr_{0-1} - 1 \geq 0.75$ (this class was measured in children ages 0–1) (From Piel et al. (2010); reproduced, with express permission, from Nature Publishing Group © 2010.)

factors could be more common in African populations because they may have been adaptive in past African environments but increase risk for disease in today's Western environment (Campbell and Tishkoff 2008).

One example of such a disease at high prevalence in African Americans is chronic kidney disease (CKD). In 2009, the mortality rate of African Americans with CKD was significantly higher than the mortality rate in European Americans and patients of other ethnicities (United States Renal Data System 2011). Additionally, in 2009, end-stage renal disease (ESRD), a serious complication of CKD in which a patient has complete or near complete loss of

renal function and requires either frequent dialysis or a kidney transplant, occurred at a rate that is 3.5 times greater in African Americans than European Americans (United States Renal Data System 2011). There are many different causes of CKD; however, the primary proximal causes include diabetes and hypertension. There are also inherited forms of CKD; some of these include polycystic kidney disease (PKD), focal segmental glomerulosclerosis (FSGS), and pediatric nephrotic syndrome (Friedman and Pollak 2011).

Most African Americans have mixed ancestry originating from Africa and Europe; studies have inferred the average amount of European ancestry in African Americans to be ~20% and the remainder to be predominantly of western and central African origin (Smith et al. 2004; Bryc et al. 2009; Tishkoff et al. 2009). It should be noted, however, that there is substantial variation in the level of African ancestry in individual African Americans. The proportion of African ancestry in a given individual can range from 99% to 1%. Additionally, the level of African or European ancestry in African Americans not only varies by individual but also by genomic region (Bryc et al. 2009). To identify genetic factors mediating disease predisposition in populations of recent African descent, like African Americans, a number of different approaches have been employed (Dong et al. 1999; Smith et al. 2004; Malhotra et al. 2005; Leak et al. 2007; Adeyemo et al. 2009; Sale et al. 2009; McDonough et al. 2011). One of these, called admixture mapping, is based on the premise that if a trait or disease of interest has a different prevalence in the parental populations prior to admixture, then it is likely that the genetic variation causing the disease or trait of interest in the admixed population is associated with chromosomal segment(s) derived from that parental population where the disease or trait is more prevalent (Winkler et al. 2010).

The identification of genetic risk factors for CKD in African Americans is an important example of the success of admixture mapping (Kao et al. 2008; Kopp et al. 2008; Genovese et al. 2010; Tzur et al. 2010). In 2008, two studies independently identified *MYH9* as a candidate locus associated with CKD or ESRD in African Americans. Kopp et al. (2008) analyzed FSGS African-American cases and controls and showed a significant association on chromosome 22 near the *MYH9* locus. They also found a marked increase of African ancestry in FSGS cases. Because *MYH9* is expressed in the kidney, the authors considered this to be a good causal candidate gene. In the second study, Kao et al. (2008) also used admixture mapping to search for genetic variants associated with CKD and identified a single location significantly associated with disease on chromosome 22 in a population of nondiabetic ESRD patients. They also showed that the amount of European ancestry

in the ESRD cases at that region was significantly lower than the average across the genome. Similar to Kopp et al. (2008), Kao et al. (2008) identified *MYH9* as the likely candidate gene responsible for the strong association signal, given the expression patterns of *MYH9* in the kidney.

Following publication of these two papers, numerous studies were conducted to find the functional variants responsible for these admixture signals (Kao et al. 2008; Kopp et al. 2008; Freedman et al. 2009a,b; Reeves-Daniel et al. 2010; Rosset et al. 2011; Freedman and Murea 2012). In 2010, two groups (Genovese et al. 2010; Tzur et al. 2010) used the data from the 1000 Genomes Project to identify potential functional variants that differ in frequency between Africans and Europeans near the *MYH9* locus. Genovese et al. (2010) performed association tests of genome-wide variants with FSGS using African-American cases and controls (a genome wide association study or GWAS). They showed that the strongest signal with FSGS was not at *MYH9* but at *APOL1*, which is a gene that is a short distance from *MYH9*. The strongest association signal was observed at a genetic variant termed G_1. When the authors controlled for the association of G_1 with FSGS they found a second associated allele that they termed G_2. The combined signal at G_1 and G_2 was 35 orders of magnitude greater than the signal found at *MYH9*. Tzur et al. (2010) also used the 1000 Genomes data to search for new variants near *MYH9* that might be functional. They found four genetic variants that were potential candidates; two variants in *APOL1* (also identified by Genovese et al. 2010), one in *APOL3*, located further upstream, and one in *FOXRED2*, located downstream from *MYH9*. They found that the *APOL1* variants are more strongly associated with CKD than the *MYH9* risk alleles in African-American and Hispanic-American ESRD cases.

In addition to the association study, Genovese et al. (2010) tested whether either the G_1 or G_2 gene variants exhibited a signature of recent positive selection. The results of their analyses suggest that at G_1 there is a strong signal of recent positive selection and a weak, but noteworthy, signal of recent positive selection at G_2.

 Cite this article as *Cold Spring Harb Perspect Biol* doi: 10.1101/cshperspect.a008524

Because it is unlikely that G_1 and G_2 were positively selected for their deleterious role in the pathogenesis of kidney disease, Genovese et al. (2010), Tzur et al. (2010), and Oleksyk et al. (2010) speculated that these alleles must serve a separate beneficial role in an African environment. In Africa, APOL1 is an important factor in resistance to infection from *Trypanosoma brucei brucei,* one of the trypanosomes that cause African sleeping sickness. Genovese et al. hypothesized that this function may be the underlying reason for the apparent recent positive selection at *APOL1* (also see Oleksyk et al. 2010; Tzur et al. 2010). To test this hypothesis, Genovese et al. (2010), examined whether the G_1 or G_2 alleles influence the resistance phenotypes conferred by APOL1 during infection caused by three subspecies of *Trypanosoma brucei* (*Trypanosoma brucei brucei, Trypanosoma brucei rhodesiense, and Trypanosoma brucei gambiens*). They showed that the G_1 and G_2 alleles effectively killed >60% of the *T.b.rhodesiense* samples and suggested that the signature of selection observed at these variants may be the result of the selective advantage conferred against *Trypanosoma* infection. However, it should be noted that *T.b.rhodesiense* is presently found in East Africa and the signature of selection identified by Genovese et al. was found in West African populations. Additionally, Tzur et al. (2010) did not find the *APOL1* variants in people from Ethiopia, which implies that the G_1 and G_2 alleles may not be common in East African populations. The absence of the G_1 and G_2 variants in Ethiopia is difficult to reconcile with the observation that that G_1 and G_2 are most effective at killing *Tryanosoma* species that are common in East Africa. However, Genovese et al. (2010) suggest that changes in *Trypanosoma* biology and distribution and/or human migration could explain the lack of correlation between parasite range and the distribution of the G_1 and G_2 alleles today. They also propose that the G_1/G_2 variants at *APOL1* could play a role in immunity to other pathogens common in western Africa.

More recently, Ko et al. (2013) further examined genetic variation at *APOL1* and showed that the G_2 variant has a similar frequency across diverse African populations (3%–8%) and that

the G_1 variant is only common in the Yoruba population (39%). These authors identified another variant, termed G_3, which has a more widespread distribution than the G_1 variant, and showed a signature of recent natural selection in a West African Fulani population. Further studies of G_1/G_2 allele frequencies in diverse African populations and additional tests of recent natural selection are necessary to better understand the evolutionary history of this locus and how selection has influenced allele frequencies at *APOL1*. It is also important to note that the functional mechanism(s) underlying the association of G_1/G_2 with CKD is currently unknown. In addition to further studies of kidney disease in African populations, additional studies are also needed to demonstrate the functional roles that the G_1, G_2, and G_3 variants play in infectious disease and kidney disease.

The identification of regions in African genomes that are recent targets of natural selection has resulted in important insights into the genetic basis of many pathological and nonpathologic phenotypes common in African populations. However, despite the growth of evolutionary studies to understand the prevalence of disease phenotypes, there is a general lack of epidemiological studies that help us understand the prevalence of noncommunicable diseases in African countries and how genetic variation plays a role in presence and prevalence of these diseases. Dalal et al. (2011) reported that in 2004 about one-quarter of all deaths in sub-Saharan Africa were caused by noncommunicable diseases, and they estimate that by 2030 noncommunicable diseases will increase to 46% of all deaths in sub-Saharan Africa. Unfortunately, their review of the literature suggests that community-based epidemiologic studies of noninfectious diseases, such as diabetes, cardiovascular disease, and obesity, are lacking in many African nations compared to the focus on maternal–child health and infectious diseases. Thus, despite the current shift in the prevalence of noninfectious diseases, we are currently ill-equipped to describe who is being affected and what the genetic and environmental risk factors that contribute to noncommunicable disease susceptibility in Africa are.

In the coming decades, it will be crucial to focus on the prevalence of noninfectious diseases in African countries and to understand what genetic variants predispose African people to conditions like obesity and cardiovascular disease. The effort to understand the genetic basis of complex noninfectious diseases in Africa should be twofold—our efforts should be placed on large-scale studies of both families and unrelated people. These two approaches will allow us to better characterize how disease-causing variants are distributed among diverse populations, and family-based studies will also help us to identify rare genetic variants that are difficult to identify in studies of unrelated people. Comparison of African populations and those within the African diaspora across diverse environments may also help to disentangle genetic and environmental effects on variable phenotypes and disease risk. These efforts will help to elucidate factors contributing to complex diseases in Africa and may also elucidate the genetic basis of these conditions in populations that have recent African admixture, like African Americans, Hispanics, and peoples of the Caribbean.

CONCLUSIONS

The pattern of genetic variation observed at mtDNA, Y chromosome, and autosomal loci in African populations reflects the demographic history of these populations. These data indicate high levels of genetic diversity within and between African populations, and also show that the current pattern of genetic variation in Africa is a result of both ancient and recent migration and admixture events.

The genomes of Africans have also been impacted by natural selection, sometimes resulting in prevalence of genetic variants that cause disease. Studies of natural selection in African populations show that functionally important genetic variants may be common but geographically restricted within Africa because of local adaptation to a particular lifestyle or environment. Also, many common variants that are adaptive because of protection from an infectious disease may also result in susceptibility to a

different, possibly noninfectious, disease in populations of recent African origin. This observation points to the importance of including ethnically diverse Africans in human genetic studies because these populations represent an important component of human genomic variation, and data from African populations help us to understand the context in which some genetic variants were selected.

Going forward, as the cost of whole genome sequencing decreases, it will become feasible to conduct large-scale genomic sequencing studies across ethnically diverse Africans. These data are necessary to understand the genetic basis of complex disease in Africa and also the genetic basis of complex disease in populations like African Americans. Currently, most large-scale studies that seek to identify genetic variants that are associated with complex disease risk are heavily biased toward the identification of genetic variants initially discovered in European populations. Large-scale sequencing studies in African populations will identify new non-European variants that can be added to the repertoire of variation that is included in large association studies. The inclusion of a more diverse panel of variants will increase what we know about the genetic basis of complex diseases because new variants will allow us to search for associations among sites that are common in people from many different ethnic backgrounds. These studies will shed light on modern human origins, African and African-American population history, and the genetic basis of nonpathologic traits as well as traits that affect disease susceptibility.

REFERENCES

Reference is also in this collection.

The 1000 Genomes Project Consortium. 2010. A map of human genome variation from population-scale sequencing. *Nature* **467:** 1061–1073.

Abi-Rached L, Jobin MJ, Kulkarni S, McWhinnie A, Dalva K, Gragert L, Babrzadeh F, Gharizadeh B, Luo M, Plummer FA, et al. 2011. The shaping of modern human immune systems by multiregional admixture with archaic humans. *Science* **334:** 89–94.

Adeyemo AA, Chen G, Chen Y, Rotimi C. 2005. Genetic structure in four West African population groups. *BMC Genet* **6:** 38.

Cite this article as *Cold Spring Harb Perspect Biol* doi: 10.1101/cshperspect.a008524

Adeyemo A, Gerry N, Chen G, Herbert A, Doumatey A, Huang H, Zhou J, Lashley K, Chen Y, Christman M, et al. 2009. A genome-wide association study of hypertension and blood pressure in African Americans. *PLoS Genet* **5**: e1000564.

Alves-Silva J, da Silva Santos M, Guimaraes PE, Ferreira AC, Bandelt HJ, Pena SD, Prado VF. 2000. The ancestry of Brazilian mtDNA lineages. *Am J Hum Genet* **67**: 444–461.

Ambrose SH. 1982. HHistorical and linguistic reconstructions in East Africa; The archeological evidence. In *The archeological and linguistic reconstruction of African History* (ed. Ehret C, Posnansky ME). Boston University African Studies Center, Berkeley, CA.

Aquadro CF, Bauer DuMont V, Reed FA. 2001. Genome-wide variation in the human and fruitfly: A comparison. *Curr Opin Genet Dev* **11**: 627–634.

Atkinson QD, Gray RD, Drummond AJ. 2008. mtDNA variation predicts population size in humans and reveals a major Southern Asian chapter in human prehistory. *Mol Biol Evol* **25**: 468–474.

Atkinson QD, Gray RD, Drummond AJ. 2009. Bayesian coalescent inference of major human mitochondrial DNA haplogroup expansions in Africa. *Proc Biol Sci* **276**: 367–373.

Atzmon G, Hao L, Pe'er I, Velez C, Pearlman A, Palamara PF, Morrow B, Friedman E, Oddoux C, Burns E, et al. 2010. Abraham's children in the genome era: Major Jewish diaspora populations comprise distinct genetic clusters with shared Middle Eastern ancestry. *Am J Hum Genet* **86**: 850–859.

Auricchio S, Rubino A, Landolt M, Semenza G, Prader A. 1963. Isolated intestinal lactase deficiency in the adult. *Lancet* **2**: 324–326.

Bandelt HJ, Forster P, Sykes BC, Richards MB. 1995. Mitochondrial portraits of human populations using median networks. *Genetics* **141**: 743–753.

Bandelt HJ, Alves-Silva J, Guimaraes PE, Santos MS, Brehm A, Pereira L, Coppa A, Larruga JM, Rengo C, Scozzari R, et al. 2001. Phylogeography of the human mitochondrial haplogroup L3e: A snapshot of African prehistory and Atlantic slave trade. *Ann Hum Genet* **65**: 549–563.

Barthelme JW. 1977. Holocene sites north-east of Lake Turkana: Preliminary report. *Azania* **12**: 33–41.

Batini C, Coia V, Battaggia C, Rocha J, Pilkington MM, Spedini G, Comas D, Destro-Bisol G, Calafell F. 2007. Phylogeography of the human mitochondrial L1c haplogroup: Genetic signatures of the prehistory of Central Africa. *Mol Phylogenet Evol* **43**: 635–644.

Batini C, Lopes J, Behar DM, Calafell F, Jorde LB, van der Veen L, Quintana-Murci L, Spedini G, Destro-Bisol G, Comas D. 2011. Insights into the demographic history of African Pygmies from complete mitochondrial genomes. *Mol Biol Evol* **28**: 1099–1110.

Batzer MA, Deininger PL. 2002. Alu repeats and human genomic diversity. *Nat Rev Genet* **3**: 370–379.

Behar DM, Villems R, Soodyall H, Blue-Smith J, Pereira L, Metspalu E, Scozzari R, Makkan H, Tzur S, Comas D, et al. 2008. The dawn of human matrilineal diversity. *Am J Hum Genet* **82**: 1130–1140.

Behar DM, Yunusbayev B, Metspalu M, Metspalu E, Rosset S, Parik J, Rootsi S, Chaubey G, Kutuev I, Yudkovsky G, et al. 2010. The genome-wide structure of the Jewish people. *Nature* **466**: 238–242.

Beleza S, Gusmao L, Amorim A, Carracedo A, Salas A. 2005. The genetic legacy of western Bantu migrations. *Hum Genet* **117**: 366–375.

Berniell-Lee G, Calafell F, Bosch E, Heyer E, Sica L, Mouguiama-Daouda P, van der Veen L, Hombert JM, Quintana-Murci L, Comas D. 2009. Genetic and demographic implications of the Bantu expansion: Insights from human paternal lineages. *Mol Biol Evol* **26**: 1581–1589.

Bersaglieri T, Sabeti PC, Patterson N, Vanderploeg T, Schaffner SF, Drake JA, Rhodes M, Reich DE, Hirschhorn JN. 2004. Genetic signatures of strong recent positive selection at the lactase gene. *Am J Hum Genet* **74**: 1111–1120.

Blench R. 1993. Recent developments in African language classification and their implications for prehistory. In *The archaeology of Africa: Food, metals and towns* (ed. Shaw PST, Andah B, Okpoko A), pp. 71–103. Routledge, London.

Blench R. 2006. *Archaeology, language, and the African past.* AltaMira, Lanham, MD.

Bower JRF. 1973. Seronera: Excavations at a stone bowl site in the Serengeti National Park, Tanzania. *Azania* **8**: 71–104.

Brancati FL, Kao WH, Folsom AR, Watson RL, Szklo M. 2000. Incident type 2 diabetes mellitus in African American and white adults: The Atherosclerosis Risk in Communities Study. *JAMA* **283**: 2253–2259.

Bray SM, Mulle JG, Dodd AF, Pulver AE, Wooding S, Warren ST. 2010. Signatures of founder effects, admixture, and selection in the Ashkenazi Jewish population. *Proc Natl Acad Sci* **107**: 16222–16227.

Bryc K, Auton A, Nelson MR, Oksenberg JR, Hauser SL, Williams S, Froment A, Bodo JM, Wambebe C, Tishkoff SA, et al. 2009. Genome-wide patterns of population structure and admixture in West Africans and African Americans. *Proc Natl Acad Sci* **107**: 786–791.

Bryc K, Auton A, Nelson MR, Oksenberg JR, Hauser SL, Williams S, Froment A, Bodo JM, Wambebe C, Tishkoff SA, et al. 2010. Genome-wide patterns of population structure and admixture in West Africans and African Americans. *Proc Natl Acad Sci* **107**: 786–791.

Butzer KW. 1981. Rise and fall of Axum, Ethiopia: A geoarchaeological interpretation. *Am. Antiq.* **46**: 471–495.

Butzer KW, Brown FH, Thurber DL. 1969. Horizontal sediments of the lower Omo Basin: The Kibish formation. *Quarternaria* **11**: 15–29.

Campbell MC, Tishkoff SA. 2008. African genetic diversity: Implications for human demographic history, modern human origins, and complex disease mapping. *Annu Rev Genomics Hum Genet* **9**: 403–433.

Cann RL, Stoneking M, Wilson AC. 1987. Mitochondrial DNA and human evolution. *Nature* **325**: 31–36.

Cann HM, de Toma C, Cazes L, Legrand MF, Morel V, Piouffre L, Bodmer J, Bodmer WF, Bonne-Tamir B, Cambon-Thomsen A, et al. 2002. A human genome diversity cell line panel. *Science* **296**: 261–262.

Castri L, Tofanelli S, Garagnani P, Bini C, Fosella X, Pelotti S, Paoli G, Pettener D, Luiselli D. 2009. mtDNA variability in two Bantu-speaking populations (Shona and Hutu) from Eastern Africa: Implications for peopling and migration patterns in sub-Saharan Africa. *Am J Phys Anthropol* **140**: 302–311.

Cavalli-Sforza LL. 1994, *The history and geography of human genes.* Princeton University Press, Princeton, NJ.

Cerny V, Hajek M, Bromova M, Cmejla R, Diallo I, Brdicka R. 2006. MtDNA of Fulani nomads and their genetic relationships to neighboring sedentary populations. *Hum Biol* **78**: 9–27.

Cerny V, Mulligan CJ, Ridl J, Zaloudkova M, Edens CM, Hajek M, Pereira L. 2008. Regional differences in the distribution of the sub-Saharan, West Eurasian, and South Asian mtDNA lineages in Yemen. *Am J Phys Anthropol* **136**: 128–137.

Cerny V, Fernandes V, Costa MD, Hajek M, Mulligan CJ, Pereira L. 2009. Migration of Chadic-speaking pastoralists within Africa based on population structure of Chad Basin and phylogeography of mitochondrial L3f haplogroup. *BMC Evol Biol* **9**: 63.

Chen YS, Torroni A, Excoffier L, Santachiara-Benerecetti AS, Wallace DC. 1995. Analysis of mtDNA variation in African populations reveals the most ancient of all human continent-specific haplogroups. *Am J Hum Genet* **57**: 133–149.

Chen YS, Olckers A, Schurr TG, Kogelnik AM, Huoponen K, Wallace DC. 2000. mtDNA variation in the South African Kung and Khwe-and their genetic relationships to other African populations. *Am J Hum Genet* **66**: 1362–1383.

Coelho M, Luiselli D, Bertorelle G, Lopes AI, Seixas S, Destro-Bisol G, Rocha J. 2005. Microsatellite variation and evolution of human lactase persistence. *Hum Genet* **117**: 329–339.

Coelho M, Sequeira F, Luiselli D, Beleza S, Rocha J. 2009. On the edge of Bantu expansions: mtDNA, Y chromosome and lactase persistence genetic variation in southwestern Angola. *BMC Evol Biol* **9**: 80.

Coia V, Destro-Bisol G, Verginelli F, Battaggia C, Boschi I, Cruciani F, Spedini G, Comas D, Calafell F. 2005. Brief communication: mtDNA variation in North Cameroon: Lack of Asian lineages and implications for back migration from Asia to sub-Saharan Africa. *Am J Phys Anthropol* **128**: 678–681.

Compton JS. 2011. Pleistocene sea-level fluctuations and human evolution on the southern coastal plain of South Africa. *Quat Sci Rev* **30**: 506–527.

* Cooper RS. 2013. Race in biological and biomedical research. *Cold Spring Harb Perspect Med* **3**: a008573.

Coudray C, Olivieri A, Achilli A, Pala M, Melhaoui M, Cherkaoui M, El-Chennawi F, Kossmann M, Torroni A, Dugoujon JM. 2008. The complex and diversified mitochondrial gene pool of Berber populations. *Ann Hum Genet* **73**: 196–214.

Cox MP, Morales DA, Woerner AE, Sozanski J, Wall JD, Hammer MF. 2009. Autosomal resequence data reveal Late Stone Age signals of population expansion in sub-Saharan African foraging and farming populations. *PLoS ONE* **4**: e6366.

Cruciani F, Santolamazza P, Shen P, Macaulay V, Moral P, Olckers A, Modiano D, Holmes S, Destro-Bisol G, Coia V, et al. 2002. A back migration from Asia to sub-Saharan

Africa is supported by high-resolution analysis of human Y-chromosome haplotypes. *Am J Hum Genet* **70**: 1197–1214.

Cruciani F, Trombetta B, Sellitto D, Massaia A, Destro-Bisol G, Watson E, Beraud Colomb E, Dugoujon JM, Moral P, Scozzari R. 2010. Human Y chromosome haplogroup R-V88: a paternal genetic record of early mid Holocene trans-Saharan connections and the spread of Chadic languages. *Eur J Hum Genet* **18**: 800–807.

Cruciani F, Trombetta B, Massaia A, Destro-Bisol G, Sellitto D, Scozzari R. 2011. A revised root for the human Y chromosomal phylogenetic tree: The origin of patrilineal diversity in Africa. *Am J Hum Genet* **88**: 814–818.

Dalal S, Beunza JJ, Volmink J, Adebamowo C, Bajunirwe F, Njelekela M, Mozaffarian D, Fawzi W, Willett W, Adami HO, et al. 2011. Non-communicable diseases in sub-Saharan Africa: What we know now. *Int J Epidemiol* **40**: 885–901.

Destro-Bisol G, Coia V, Boschi I, Verginelli F, Caglia A, Pascali V, Spedini G, Calafell F. 2004. The analysis of variation of mtDNA hypervariable region 1 suggests that Eastern and Western Pygmies diverged before the Bantu expansion. *Am Nat* **163**: 212–226.

de Wit E, Delport W, Rugamika CE, Meintjes A, Moller M, van Helden PD, Seoighe C, Hoal EG. 2010. Genome-wide analysis of the structure of the South African coloured population in the Western Cape. *Hum Genet* **128**: 145–153.

Distefano JA. 1990. Hunters or hunted? Towards a history of the Okiek of Kenya. *Hist Africa* **17**: 41–57.

Dong Y, Zhu H, Sagnella GA, Carter ND, Cook DG, Cappuccio FP. 1999. Association between the C825T polymorphism of the G protein beta3-subunit gene and hypertension in blacks. *Hypertension* **34**: 1193–1196.

Durham WH. 1991. *Coevolution. Genes, culture and human diversity.* Stanford University Press, Stanford, CT.

Ehret C. 1967. Cattle-keeping and milking in Eastern and Southern African History: The linguistic evidence. *J African Hist* **8**: 1–17.

Ehret C. 1971. *Southern Nilotic history: Linguistic approaches to the study of the past.* Northwestern University Press, Evanston, IL.

Ehret C. 1993. Nilo-Saharans and the Saharo-Sudanese Neolithic. In *The archaeology of Africa: Food, metals and towns* (ed. Shaw PST, Andah B, Okpoko A), pp. 104–125. Routledge, London.

Ehret C. 1995. *Reconstructing proto-Afroasiatic (Proto-Afrasian): Vowels, tone, consonants, and vocabulary.* University of California Press, Berkeley, CA.

Ehret C. 1998. *An African classical age: Eastern and Southern Africa in world history, 1000 B.C. to A.D. 400.* University Press of Virginia, Charlottesville, VA.

Ehret C. 2001a. Bantu expansions: Re-envisioning a central problem of early African History. *Int J African Historic Stud* **34**: 5–41.

Ehret C. 2001b. *Historical-comparative reconstruction of Nilo-Saharan.* Rüdiger Köppe Verlag, Cologne, Germany.

Ehret C. 2002. *The civilizations of Africa: A history to 1800.* University Press of Virginia, Charlottesville, VA.

Ehret C. 2006. Linguistic stratigraphies and Holocene history in Northeastern Africa. In *Studies in African archae-*

Cite this article as *Cold Spring Harb Perspect Biol* doi: 10.1101/cshperspect.a008524

ology, pp. 1019–1055. Poznan Archaeological Museum, Poznan, Poland.

Enattah NS, Sahi T, Savilahti E, Terwilliger JD, Peltonen L, Jarvela I. 2002. Identification of a variant associated with adult-type hypolactasia. *Nat Genet* **30:** 233–237.

Enattah NS, Jensen TG, Nielsen M, Lewinski R, Kuokkanen M, Rasinpera H, El-Shanti H, Seo JK, Alifrangis M, Khalil IF, et al. 2008. Independent introduction of two lactase-persistence alleles into human populations reflects different history of adaptation to milk culture. *Am J Hum Genet* **82:** 57–72.

Evershed RP, Payne S, Sherratt AG, Copley MS, Coolidge J, Urem-Kotsu D, Kotsakis K, Ozdogan M, Ozdogan AE, Nieuwenhuyse O, et al. 2008. Earliest date for milk use in the Near East and southeastern Europe linked to cattle herding. *Nature* **455:** 528–531.

Excoffier L, Schneider S. 1999. Why hunter–gatherer populations do not show signs of pleistocene demographic expansions. *Proc Natl Acad Sci* **96:** 10597–10602.

Flint J, Hill AV, Bowden DK, Oppenheimer SJ, Sill PR, Serjeantson SW, Bana-Koiri J, Bhatia K, Alpers MP, Boyce AJ, et al. 1986. High frequencies of α-thalassaemia are the result of natural selection by malaria. *Nature* **321:** 744–750.

Fortin A, Stevenson MM, Gros P. 2002. Susceptibility to malaria as a complex trait: Big pressure from a tiny creature. *Hum Mol Genet* **11:** 2469–2478.

Francalacci P, Morelli L, Angius A, Berutti R, Reinier F, Atzeni R, Pilu R, Busonero F, Maschio A, Zara I, et al. 2013. Low-pass DNA sequencing of 1200 Sardinians reconstructs European Y-chromosome phylogeny. *Science* **341:** 565–569.

Freedman BI, Murea M. 2012. Target organ damage in African American hypertension: Role of APOL1. *Curr Hypertens Rep* **14:** 21–28.

Friedman DJ, Pollak MR. 2011. Genetics of kidney failure and the evolving story of APOL1. *J Clin Invest* **121:** 3367–3374.

Freedman BI, Hicks PJ, Bostrom MA, Cunningham ME, Liu Y, Divers J, Kopp JB, Winkler CA, Nelson GW, Langefeld CD, et al. 2009a. Polymorphisms in the non-muscle myosin heavy chain 9 gene (MYH9) are strongly associated with end-stage renal disease historically attributed to hypertension in African Americans. *Kidney Int* **75:** 736–745.

Freedman BI, Kopp JB, Winkler CA, Nelson GW, Rao DC, Eckfeldt JH, Leppert MF, Hicks PJ, Divers J, Langefeld CD, et al. 2009b. Polymorphisms in the nonmuscle myosin heavy chain 9 gene (MYH9) are associated with albuminuria in hypertensive African Americans: The HyperGEN study. *Am J Nephrol* **29:** 626–632.

Genovese G, Friedman DJ, Ross MD, Lecordier L, Uzureau P, Freedman BI, Bowden DW, Langefeld CD, Oleksyk TK, Uscinski Knob AL, et al. 2010. Association of trypanolytic ApoL1 variants with kidney disease in African Americans. *Science* **329:** 841–845.

Gomez F, Ko WY, Davis A, Tishkoff SA. 2013a. Malarial disease and human genetic variation: Evidence for natural selection at malaria susceptibility candidate loci. In *Primates, pathogens and evolution* (ed. Brinkworth J, Pechenkina E). Springer, New York.

Gomez F, Tomas G, Ko WY, Ranciaro A, Froment A, Ibrahim M, Lema G, Nyambo TB, Omar SA, Wambebe C, et al. 2013b. Patterns of nucleotide and haplotype diversity at ICAM-1 across global human populations with varying levels of malaria exposure. *Hum Genet* **132:** 987–999.

Gonder MK, Mortensen HM, Reed FA, de Sousa A, Tishkoff SA. 2007. Whole-mtDNA genome sequence analysis of ancient African lineages. *Mol Biol Evol* **24:** 757–768.

Gonzalez AM, Cabrera VM, Larruga JM, Tounkara A, Noumsi G, Thomas BN, Moulds JM. 2006. Mitochondrial DNA variation in Mauritania and Mali and their genetic relationship to other Western Africa populations. *Ann Hum Genet* **70:** 631–657.

Graven L, Passarino G, Semino O, Boursot P, Santachiara-Benerecetti S, Langaney A, Excoffier L. 1995. Evolutionary correlation between control region sequence and restriction polymorphisms in the mitochondrial genome of a large Senegalese Mandenka sample. *Mol Biol Evol* **12:** 334–345.

Green RE, Krause J, Briggs AW, Maricic T, Stenzel U, Kircher M, Patterson N, Li H, Zhai W, Fritz MH-Y, et al. 2011. A draft sequence of the Neandertal genome. *Science* **328:** 710–722.

Greenberg JH. 1963. *The languages of Africa*. Indiana University, Bloomington.

Greenberg JH. 1972. Linguistic evidence regarding Bantu origins. *J Afr Hist* **13:** 189–216.

Haldane JBS. 1949. The rate of mutation in human genes. *Hereditas* **35:** 267–273.

Hammer MF, Zegura SL. 2002. The human Y Chromosome haplogroup tree: Nomenclature and phylogeography of its major divisions. *Annu Rev Anthropol* **31:** 303–321.

Hammer MF, Woerner AE, Mendez FL, Watkins JC, Wall JD. 2011. Genetic evidence for archaic admixture in Africa. *Proc Natl Acad Sci* **108:** 15123–15128.

Harpending HC, Sherry ST, Rogers AR, Stoneking M. 1993. The genetic structure of ancient human populations. *Curr Anthropol* **34:** 483–496.

Haspelmath M, Dryer MS, Gil D, Comrie B, eds. 2008. *The world atlas of language structures online*. Max Planck Digital Library, Munich.

Henn BM, Gignoux CR, Jobin M, Granka JM, Macpherson JM, Kidd JM, Rodriguez-Botigue L, Ramachandran S, Hon L, Brisbin A, et al. 2011. From the cover: Feature article: Hunter–gatherer genomic diversity suggests a southern African origin for modern humans. *Proc Natl Acad Sci* **108:** 5154–5162.

Henn BM, Botigué LR, Gravel S, Wang W, Brisbin A, Byrnes JK, Fadhlaoui-Zid K, Zalloua PA, Moreno-Estrada A, Bertranpetit J, et al. 2012. Genomic ancestry of North Africans supports back-to-Africa migrations. *PLoS Genet* **8:** e1002397.

Herrnstadt C, Elson JL, Fahy E, Preston G, Turnbull DM, Anderson C, Ghosh SS, Olefsky JM, Beal MF, Davis RE, et al. 2002. Reduced-median-network analysis of complete mitochondrial DNA coding-region sequences for the major African, Asian, and European haplogroups. *Am J Hum Genet* **70:** 1152–1171.

Hofer T, Ray N, Wegmann D, Excoffier L. 2009. Large allele frequency differences between human continental groups are more likely to have occurred by drift during range expansions than by selection. *Ann Hum Genet* **73:** 95–108.

Hunter DJ. 2005. Gene-environment interactions in human diseases. *Nat Rev Genet* **6:** 287–298.

Hunter-Zinck H, Musharoff S, Salit J, Al-Ali KA, Chouchane L, Gohar A, Matthews R, Butler MW, Fuller J, Hackett NR, et al. 2010. Population genetic structure of the people of Qatar. *Am J Hum Genet* **87:** 17–25.

Ingman M, Kaessmann H, Paabo S, Gyllensten U. 2000. Mitochondrial genome variation and the origin of modern humans. *Nature* **408:** 708–713.

Ingram CJ, Elamin MF, Mulcare CA, Weale ME, Tarekegn A, Raga TO, Bekele E, Elamin FM, Thomas MG, Bradman N, et al. 2007. A novel polymorphism associated with lactose tolerance in Africa: Multiple causes for lactase persistence? *Hum Genet* **120:** 779–788.

Ingram CJ, Mulcare CA, Itan Y, Thomas MG, Swallow DM. 2009. Lactose digestion and the evolutionary genetics of lactase persistence. *Hum Genet* **124:** 579–591.

International HapMap C. 2005. A haplotype map of the human genome. *Nature* **437:** 1299–1320.

Jackson BA, Wilson JL, Kirbah S, Sidney SS, Rosenberger J, Bassie L, Alie JA, McLean DC, Garvey WT, Ely B. 2005. Mitochondrial DNA genetic diversity among four ethnic groups in Sierra Leone. *Am J Phys Anthropol* **128:** 156–163.

Kao WH, Klag MJ, Meoni LA, Reich D, Berthier-Schaad Y, Li M, Coresh J, Patterson N, Tandon A, Powe NR, et al. 2008. MYH9 is associated with nondiabetic end-stage renal disease in African Americans. *Nat Genet* **40:** 1185–1192.

Karafet TM, Mendez FL, Meilerman MB, Underhill PA, Zegura SL, Hammer MF. 2008. New binary polymorphisms reshape and increase resolution of the human Y chromosomal haplogroup tree. *Genome Res* **18:** 830–838.

Kivisild T, Reidla M, Metspalu E, Rosa A, Brehm A, Pennarun E, Parik J, Geberhiwot T, Usanga E, Villems R. 2004. Ethiopian mitochondrial DNA heritage: Tracking gene flow across and around the gate of tears. *Am J Hum Genet* **75:** 752–770.

Ko WY, Gomez F, Tishkoff S. 2012. Evolution of human erythrocyte-specific genes involved in malaria susceptibility. In *Rapidly evolving genes and genetic systems* (ed. Singh RS, Xu J, Kulathinal RJ). Oxford University Press, Oxford.

Kopelman NM, Stone L, Wang C, Gefel D, Feldman MW, Hillel J, Rosenberg NA. 2009. Genomic microsatellites identify shared Jewish ancestry intermediate between Middle Eastern and European populations. *BMC Genet* **10:** 80.

Kopp JB, Smith MW, Nelson GW, Johnson RC, Freedman BI, Bowden DW, Oleksyk T, McKenzie LM, Kajiyama H, Ahuja TS, et al. 2008. MYH9 is a major-effect risk gene for focal segmental glomerulosclerosis. *Nat Genet* **40:** 1175–1184.

Krings M, Salem AE, Bauer K, Geisert H, Malek AK, Chaix L, Simon C, Welsby D, Di Rienzo A, Utermann G, et al. 1999. mtDNA analysis of Nile River Valley populations: A genetic corridor or a barrier to migration? *Am J Hum Genet* **64:** 1166–1176.

Labuda D, Zietkiewicz E, Yotova V. 2000. Archaic lineages in the history of modern humans. *Genetics* **156:** 799–808.

Lachance J, Vernot B, Elbers CC, Ferwerda B, Froment A, Bodo JM, Lema G, Fu W, Nyambo TB, Rebbeck TR, et al. 2012. Evolutionary history and adaptation from high-coverage whole-genome sequences of diverse African hunter–gatherers. *Cell* **150:** 457–469.

Landis SH, Murray T, Bolden S, Wingo PA. 1999. Cancer statistics, 1999. *CA Cancer J Clin* **49:** 8–31.

Leak TS, Keene KL, Langefeld CD, Gallagher CJ, Mychaleckyj JC, Freedman BI, Bowden DW, Rich SS, Sale MM. 2007. Association of the proprotein convertase subtilisin/kexin-type 2 (PCSK2) gene with type 2 diabetes in an African American population. *Mol Genet Metab* **92:** 145–150.

Leakey LSB. 1931. *The Stone Age cultures of Kenya Colony.* The University Press, Cambridge.

Lewinsky RH, Jensen TG, Moller J, Stensballe A, Olsen J, Troelsen JT. 2005. T-13910 DNA variant associated with lactase persistence interacts with Oct-1 and stimulates lactase promoter activity in vitro. *Hum Mol Genet* **14:** 3945–3953.

Li JZ, Absher DM, Tang H, Southwick AM, Casto AM, Ramachandran S, Cann HM, Barsh GS, Feldman M, Cavalli-Sforza LL, et al. 2008. Worldwide human relationships inferred from genome-wide patterns of variation. *Science* **319:** 1100–1104.

Luis JR, Rowold DJ, Regueiro M, Caeiro B, Cinnioglu C, Roseman C, Underhill PA, Cavalli-Sforza LL, Herrera RJ. 2004. The Levant versus the Horn of Africa: Evidence for bidirectional corridors of human migrations. *Am J Hum Genet* **74:** 532–544.

Maca-Meyer N, Gonzalez AM, Larruga JM, Flores C, Cabrera VM. 2001. Major genomic mitochondrial lineages delineate early human expansions. *BMC Genet* **2:** 13.

Macaulay V, Richards M, Hickey E, Vega E, Cruciani F, Guida V, Scozzari R, Batsheva B-T, Sykes B, Torroni A. 1999. The emerging tree of West Eurasian mtDNAs: A synthesis of control region sequences and RFLPs. *Am J Hum Genet* **64:** 232–249.

Malhotra A, Coon H, Feitosa MF, Li WD, North KE, Price RA, Bouchard C, Hunt SC, Wolford JK. 2005. Meta-analysis of genome-wide linkage studies for quantitative lipid traits in African Americans. *Hum Mol Genet* **14:** 3955–3962.

Marth GT, Czabarka E, Murvai J, Sherry ST. 2004. The allele frequency spectrum in genome-wide human variation data reveals signals of differential demographic history in three large world populations. *Genetics* **166:** 351–372.

McDonough CW, Palmer ND, Hicks PJ, Roh BH, An SS, Cooke JN, Hester JM, Wing MR, Bostrom MA, Rudock ME, et al. 2011. A genome-wide association study for diabetic nephropathy genes in African Americans. *Kidney Int* **79:** 563–572.

Mendez Fernando L, Krahn T, Schrack B, Krahn A-M, Veeramah Krishna R, Woerner August E, Fomine Forka Leypey M, Bradman N, Thomas Mark G, Karafet Tatiana M, et al. 2013. An African American paternal lineage adds an extremely ancient root to the human Y chromosome phylogenetic tree. *Am J Hum Genet* **92:** 454–459.

Mishmar D, Ruiz-Pesini E, Golik P, Macaulay V, Clark AG, Hosseini S, Brandon M, Easley K, Chen E, Brown MD, et al. 2003. Natural selection shaped regional mtDNA variation in humans. *Proc Natl Acad Sci* **100:** 171–176.

Cite this article as *Cold Spring Harb Perspect Biol* doi: 10.1101/cshperspect.a008524

Mulcare CA, Weale ME, Jones AL, Connell B, Zeitlyn D, Tarekegn A, Swallow DM, Bradman N, Thomas MG. 2004. The T allele of a single-nucleotide polymorphism 13.9 kb upstream of the lactase gene (LCT) (C-13.9kbT) does not predict or cause the lactase-persistence phenotype in Africans. *Am J Hum Genet* **74:** 1102–1110.

Newcomer AD, Park HS, O'Brien PC, McGill DB. 1983. Response of patients with irritable bowel syndrome and lactase deficiency using unfermented acidophilus milk. *Am J Clin Nutr* **38:** 257–263.

Newman P. 1980. *The classification of Chadic within Afroasiatic.* Leiden Universitaire Press, Leiden.

Nielsen R. 2005. Molecular signatures of natural selection. *Annu Rev Genet* **39:** 197–218.

Nurse D. 1997. The contributions of linguistics to the study of history in Africa. *Journal Afr Hist* **38:** 359–391.

Olds LC, Sibley E. 2003. Lactase persistence DNA variant enhances lactase promoter activity in vitro: Functional role as a *cis* regulatory element. *Hum Mol Genet* **12:** 2333–2340.

Oleksyk TK, Nelson GW, An P, Kopp JB, Winkler CA. 2010. Worldwide distribution of the MYH9 kidney disease susceptibility alleles and haplotypes: Evidence of historical selection in Africa. *PLoS ONE* **5:** e11474.

Ong KL, Cheung BM, Man YB, Lau CP, Lam KS. 2007. Prevalence, awareness, treatment, and control of hypertension among United States adults 1999–2004. *Hypertension* **49:** 69–75.

Pagani L, Kivisild T, Tarekegn A, Ekong R, Plaster C, Gallego Romero I, Ayub Q, Mehdi SQ, Thomas MG, Luiselli D, et al. 2012. Ethiopian genetic diversity reveals linguistic stratification and complex influences on the Ethiopian gene pool. *Am J Hum Genet* **91:** 83–96.

Patterson N, Petersen DC, van der Ross RE, Sudoyo H, Glashoff RH, Marzuki S, Reich D, Hayes VM. 2010. Genetic structure of a unique admixed population: Implications for medical research. *Hum Mol Genet* **19:** 411–419.

Pereira L, Macaulay V, Torroni A, Scozzari R, Prata MJ, Amorim A. 2001. Prehistoric and historic traces in the mtDNA of Mozambique: Insights into the Bantu expansions and the slave trade. *Ann Hum Genet* **65:** 439–458.

Phillipson L. 2009. Lithic artefacts as a source of cultural, social and economic information: The evidence from Aksum, Ethiopia. *Afr Archaeol Rev* **26:** 45–58.

Piel FB, Patil AP, Howes RE, Nyangiri OA, Gething PW, Williams TN, Weatherall DJ, Hay SI. 2010. Global distribution of the sickle cell gene and geographical confirmation of the malaria hypothesis. *Nat Commun* **1:** 104.

Plagnol V, Wall JD. 2006. Possible ancestral structure in human populations. *PLoS Genet* **2:** e105.

Poloni ES, Naciri Y, Bucho R, Niba R, Kervaire B, Excoffier L, Langaney A, Sanchez-Mazas A. 2009. Genetic evidence for complexity in ethnic differentiation and history in East Africa. *Ann Hum Genet* **73:** 582–600.

Posnansky M. 1961a. 168. Dimple-based pottery from Uganda. *Man* **61:** 141–142.

Posnansky M. 1961b. Pottery types from archaeological sites in East Africa. *J Afr Hist* **2:** 177–198.

Poulter M, Hollox E, Harvey CB, Mulcare C, Peuhkuri K, Kajander K, Sarner M, Korpela R, Swallow DM. 2003.

The causal element for the lactase persistence/non-persistence polymorphism is located in a 1 Mb region of linkage disequilibrium in Europeans. *Ann Hum Genet* **67:** 298–311.

Poznik GD, Henn BM, Yee M-C, Sliwerska E, Euskirchen GM, Lin AA, Snyder M, Quintana-Murci L, Kidd JM, Underhill PA, et al. 2013. Sequencing Y chromosomes resolves discrepancy in time to common ancestor of males versus females. *Science* **341:** 562–565.

Prugnolle F, Manica A, Balloux F. 2005. Geography predicts neutral genetic diversity of human populations. *Curr Biol* **15:** R159–160.

Quintana-Murci L, Semino O, Bandelt HJ, Passarino G, McElreavey K, Santachiara-Benerecetti AS. 1999. Genetic evidence of an early exit of *Homo sapiens sapiens* from Africa through eastern Africa. *Nat Genet* **23:** 437–441.

Quintana-Murci L, Quach H, Harmant C, Luca F, Massonnet B, Patin E, Sica L, Mouguiama-Daouda P, Comas D, Tzur S, et al. 2008. Maternal traces of deep common ancestry and asymmetric gene flow between Pygmy hunter–gatherers and Bantu-speaking farmers. *Proc Natl Acad Sci* **105:** 1596–1601.

Quintana-Murci L, Harmant C, Quach H, Balanovsky O, Zaporozhchenko V, Bormans C, van Helden PD, Hoal EG, Behar DM. 2010. Strong maternal Khoisan contribution to the South African coloured population: A case of gender-biased admixture. *Am J Hum Genet* **86:** 611–620.

Ramachandran S, Deshpande O, Roseman CC, Rosenberg NA, Feldman MW, Cavalli-Sforza LL. 2005. Support from the relationship of genetic and geographic distance in human populations for a serial founder effect originating in Africa. *Proc Natl Acad Sci* **102:** 15942–15947.

Rando JC, Pinto F, Gonzalez AM, Hernandez M, Larruga JM, Cabrera VM, Bandelt HJ. 1998. Mitochondrial DNA analysis of northwest African populations reveals genetic exchanges with European, near-eastern, and sub-Saharan populations. *Ann Hum Genet* **62:** 531–550.

Ray N, Currat M, Berthier P, Excoffier L. 2005. Recovering the geographic origin of early modern humans by realistic and spatially explicit simulations. *Genome Res* **15:** 1161–1167.

Reeves-Daniel AM, Iskandar SS, Bowden DW, Bostrom MA, Hicks PJ, Comeau ME, Langefeld CD, Freedman BI. 2010. Is collapsing C1q nephropathy another MYH9-associated kidney disease? A case report. *Am J Kidney Dis* **55:** e21–24.

Reich D, Green RE, Kircher M, Krause J, Patterson N, Durand EY, Viola B, Briggs AW, Stenzel U, Johnson PLF, et al. 2011. Genetic history of an archaic hominin group from Denisova Cave in Siberia. *Nature* **468:** 1053–1060.

Richards MB, Macaulay VA, Bandelt HJ, Sykes BC. 1998. Phylogeography of mitochondrial DNA in western Europe. *Ann Hum Genet* **62:** 241–260.

Richards M, Macaulay V, Hickey E, Vega E, Sykes B, Guida V, Rengo C, Sellitto D, Cruciani F, Kivisild T, et al. 2000. Tracing European founder lineages in the Near Eastern mtDNA pool. *Am J Hum Genet* **67:** 1251–1276.

Robbins LH. 1972. Archeology in the Turkana District, Kenya. *Science* **176:** 359–366.

Rogers AR, Harpending H. 1992. Population growth makes waves in the distribution of pairwise genetic differences. *Mol Biol Evol* **9:** 552–569.

Ronald J, Akey JM. 2005. Genome-wide scans for loci under selection in humans. *Hum Genomics* **2:** 113–125.

Rosenberg NA, Pritchard JK, Weber JL, Cann HM, Kidd KK, Zhivotovsky LA, Feldman MW. 2002. Genetic structure of human populations. *Science* **298:** 2381–2385.

Rosset S, Tzur S, Behar DM, Wasser WG, Skorecki K. 2011. The population genetics of chronic kidney disease: Insights from the MYH9-APOL1 locus. *Nat Rev Nephrol* **7:** 313–326.

Sabeti PC, Reich DE, Higgins JM, Levine HZ, Richter DJ, Schaffner SF, Gabriel SB, Platko JV, Patterson NJ, McDonald GJ, et al. 2002. Detecting recent positive selection in the human genome from haplotype structure. *Nature* **419:** 832–837.

Sabeti PC, Schaffner SF, Fry B, Lohmueller J, Varilly P, Shamovsky O, Palma A, Mikkelsen TS, Altshuler D, Lander ES. 2006. Positive natural selection in the human lineage. *Science* **312:** 1614–1620.

Salas A, Richards M, De la Fe T, Lareu MV, Sobrino B, Sanchez-Diz P, Macaulay V, Carracedo A. 2002. The making of the African mtDNA landscape. *Am J Hum Genet* **71:** 1082–1111.

Sale MM, Lu L, Spruill IJ, Fernandes JK, Lok KH, Divers J, Langefeld CD, Garvey WT. 2009. Genome-wide linkage scan in Gullah-speaking African American families with type 2 diabetes: The Sea Islands Genetic African American Registry (Project SuGAR). *Diabetes* **58:** 260–267.

Satta Y, Takahata N. 2004. The distribution of the ancestral haplotype in finite stepping-stone models with population expansion. *Mol Ecol* **13:** 877–886.

Saunders MA, Hammer MF, Nachman MW. 2002. Nucleotide variability at G6pd and the signature of malarial selection in humans. *Genetics* **162:** 1849–1861.

Saunier JL, Irwin JA, Strouss KM, Ragab H, Sturk KA, Parsons TJ. 2009. Mitochondrial control region sequences from an Egyptian population sample. *Forensic Sci Int Genet* **3:** e97–103.

Schlebusch CM, Skoglund P, Sjodin P, Gattepaille LM, Hernandez D, Jay F, Li S, De Jongh M, Singleton A, Blum MG, et al. 2012. Genomic variation in seven Khoe-San groups reveals adaptation and complex African history. *Science* **338:** 374–379.

Scozzari R, Cruciani F, Santolamazza P, Malaspina P, Torroni A, Sellitto D, Arredi B, Destro-Bisol G, De Stefano G, Rickards O, et al. 1999. Combined use of biallelic and microsatellite Y-chromosome polymorphisms to infer affinities among African populations. *Am J Hum Genet* **65:** 829–846.

Segal I, Gagjee PP, Essop AR, Noormohamed AM. 1983. Lactase deficiency in the South African black population. *Am J Clin Nutr* **38:** 901–905.

Seielstad MT, Minch E, Cavalli-Sforza LL. 1998. Genetic evidence for a higher female migration rate in humans. *Nat Genet* **20:** 278–280.

Sikora M, Laayouni H, Calafell F, Comas D, Bertranpetit J. 2010. A genomic analysis identifies a novel component in the genetic structure of sub-Saharan African populations. *Eur J Hum Genet* **19:** 84–88.

Smith AB. 1992. Origins and spread of pastoralism in Africa. *Annu Rev Anthropol* **21:** 125–141.

Smith MW, Patterson N, Lautenberger JA, Truelove AL, McDonald GJ, Waliszewska A, Kessing BD, Malasky MJ, Scafe C, Le E, et al. 2004. A high-density admixture map for disease gene discovery in African Americans. *Am J Hum Genet* **74:** 1001–1013.

Soodyall H, Vigilant L, Hill AV, Stoneking M, Jenkins T. 1996. mtDNA control-region sequence variation suggests multiple independent origins of an "Asian-specific" 9-bp deletion in sub-Saharan Africans. *Am J Hum Genet* **58:** 595–608.

Stefflova K, Dulik MC, Pai AA, Walker AH, Zeigler-Johnson CM, Gueye SM, Schurr TG, Rebbeck TR. 2009. Evaluation of group genetic ancestry of populations from Philadelphia and Dakar in the context of sex-biased admixture in the Americas. *PLoS ONE* **4:** e7842.

Stringer C. 1994. Out of Africa—A personal history. In *Origins of anatomically modern humans* (ed. Nitecki M, Nitecki D). Plenum, New York.

Stringer CB, Andrews P. 1988. Genetic and fossil evidence for the origin of modern humans. *Science* **239:** 1263–1268.

Sutton J. 1973. The settlement of East Africa. In *Zamani* (ed. Ogot BA), pp. 70–97. Longmans, Nairobi.

Swallow DM. 2003. Genetics of lactase persistence and lactose intolerance. *Annu Rev Genet* **37:** 197–219.

Templeton A. 2002. Out of Africa again and again. *Nature* **416:** 45–51.

Thomas MG, Weale ME, Jones AL, Richards M, Smith A, Redhead N, Torroni A, Scozzari R, Gratrix F, Tarekegn A, et al. 2002. Founding mothers of Jewish communities: Geographically separated Jewish groups were independently founded by very few female ancestors. *Am J Hum Genet* **70:** 1411–1420.

Thomason SGaK, T. 1988. *Language contact, creolization, and genetic linguistics*. University of California Press, Berkeley and Los Angeles.

Tishkoff SA, Goldman A, Calafell F, Speed WC, Deinard AS, Bonne-Tamir B, Kidd JR, Pakstis AJ, Jenkins T, Kidd KK. 1998. A global haplotype analysis of the myotonic dystrophy locus: Implications for the evolution of modern humans and for the origin of myotonic dystrophy mutations. *Am J Hum Genet* **62:** 1389–1402.

Tishkoff SA, Varkonyi R, Cahinhinan N, Abbes S, Argyropoulos G, Destro-Bisol G, Drousiotou A, Dangerfield B, Lefranc G, Loiselet J, et al. 2001. Haplotype diversity and linkage disequilibrium at human G6PD: Recent origin of alleles that confer malarial resistance. *Science* **293:** 455–462.

Tishkoff SA, Reed FA, Ranciaro A, Voight BF, Babbitt CC, Silverman JS, Powell K, Mortensen HM, Hirbo JB, Osman M, et al. 2007. Convergent adaptation of human lactase persistence in Africa and Europe. *Nat Genet* **39:** 31–40.

Tishkoff SA, Reed FA, Friedlaender FR, Ehret C, Ranciaro A, Froment A, Hirbo JB, Awomoyi AA, Bodo J-M, Doumbo O, et al. 2009. The genetic structure and history of Africans and African Americans. *Science* **324:** 1035–1044.

Torroni A, Rengo C, Guida V, Cruciani F, Sellitto D, Coppa A, Calderon FL, Simionati B, Valle G, Richards M, et al.

Cite this article as *Cold Spring Harb Perspect Biol* doi: 10.1101/cshperspect.a008524

2001. Do the four clades of the mtDNA haplogroup L2 evolve at different rates? *Am J Hum Genet* **69:** 1348–1356.

Troelsen JT, Olsen J, Moller J, Sjostrom H. 2003. An upstream polymorphism associated with lactase persistence has increased enhancer activity. *Gastroenterology* **125:** 1686–1694.

Tzur S, Rosset S, Shemer R, Yudkovsky G, Selig S, Tarekegn A, Bekele E, Bradman N, Wasser WG, Behar DM, et al. 2010. Missense mutations in the APOL1 gene are highly associated with end stage kidney disease risk previously attributed to the MYH9 gene. *Hum Genet* **128:** 345–350.

United States Renal Data System 2011. *USRDS 2011 annual data report: Atlas of chronic kidney disease and end-stage renal disease in the United States.* National Institute of Diabetes and Digestive and Kidney Diseases, Bethesda, MD.

Van Beek GW. 1967. Monuments of Axum in the light of South Arabian archeology. *J Am Orient Soc* **87:** 113–122.

Vansina J. 1995. New linguistic evidence and "the Bantu Expansion." *J Afr His* **36:** 173–195.

Veeramah KR, Connell BA, Pour NA, Powell A, Plaster CA, Zeitlyn D, Mendell NR, Weale ME, Bradman N, Thomas MG. 2010. Little genetic differentiation as assessed by uniparental markers in the presence of substantial language variation in peoples of the Cross River region of Nigeria. *BMC Evol Biol* **10:** 92.

Verrelli BC, McDonald JH, Argyropoulos G, Destro-Bisol G, Froment A, Drousiotou A, Lefranc G, Helal AN, Loiselet J, Tishkoff SA. 2002. Evidence for balancing selection from nucleotide sequence analyses of human G6PD. *Am J Hum Genet* **71:** 1112–1128.

Vigilant L. 1990. Control region sequences from African populations and the evolution of human mitochondrial DNA. University of California, Berkeley, CA.

Vigilant L, Stoneking M, Harpending H, Hawkes K, Wilson AC. 1991. African populations and the evolution of human mitochondrial DNA. *Science* **253:** 1503–1507.

Wall JD, Lohmueller KE, Plagnol V. 2009. Detecting ancient admixture and estimating demographic parameters in multiple human populations. *Mol Biol Evol* **26:** 1823–1827.

Wallace DC, Brown MD, Lott MT. 1999. Mitochondrial DNA variation in human evolution and disease. *Gene* **238:** 211–230.

Wang S, Lachance J, Tishkoff SA, Hey J, Xing J. 2013. Apparent variation in Neanderthal admixture among African populations is consistent with gene flow from non-African populations. *Genome Biol Evol* **5:** 2075–2081.

Watson E, Bauer K, Aman R, Weiss G, von Haeseler A, Paabo S. 1996. mtDNA sequence diversity in Africa. *Am J Hum Genet* **59:** 437–444.

Watson E, Forster P, Richards M, Bandelt HJ. 1997. Mitochondrial footprints of human expansion in Africa. *Am J Hum Genet* **61:** 691–704.

Weatherall DJ. 2001. Phenotype-genotype relationships in monogenic disease: Lessons from the thalassaemias. *Nat Rev Genet* **2:** 245–255.

Wei W, Ayub Q, Chen Y, McCarthy S, Hou Y, Carbone I, Xue Y, Tyler-Smith C. 2013. A calibrated human Y-chromosomal phylogeny based on resequencing. *Genome Res* **23:** 388–395.

Winkler CA, Nelson GW, Smith MW. 2010. Admixture mapping comes of age. *Annu Rev Genomics Hum Genet* **11:** 65–89.

Wood ET, Stover DA, Ehret C, Destro-Bisol G, Spedini G, McLeod H, Louie L, Bamshad M, Strassmann BI, Soodyall H, et al. 2005. Contrasting patterns of Y chromosome and mtDNA variation in Africa: Evidence for sex-biased demographic processes. *Eur J Hum Genet* **13:** 867–876.

Yang Q, Khoury MJ, Friedman J, Little J, Flanders WD. 2005. How many genes underlie the occurrence of common complex diseases in the population? *Int J Epidemiol* **34:** 1129–1137.

Yotova V, Lefebvre J-F, Moreau C, Gbeha E, Hovhannesyan K, Bourgeois S, Bédarida S, Azevedo L, Amorim A, Sarkisian T, et al. 2011. An X-linked haplotype of Neandertal origin is present among all non-African populations. *Mol Biol Evol* **28:** 1957–1962.

Zietkiewicz E, Yotova V, Jarnik M, Korab-Laskowska M, Kidd KK, Modiano D, Scozzari R, Stoneking M, Tishkoff S, Batzer M, et al. 1998. Genetic structure of the ancestral population of modern humans. *J Mol Evol* **47:** 146–155.

A Genomic View of the Peopling and Population Structure of India

Partha P. Majumder and Analabha Basu

National Institute of Biomedical Genomics, Kalyani 741251, India

Correspondence: ppm1@nibmg.ac.in

Recent advances in molecular and statistical genetics have enabled the reconstruction of human history by studying living humans. The ability to sequence and study DNA by calibrating the rate of accumulation of changes with evolutionary time has enabled robust inferences about how humans have evolved. These data indicate that modern humans evolved in Africa about 150,000 years ago and, consistent with paleontological evidence, migrated out of Africa. And through a series of settlements, demographic expansions, and further migrations, they populated the entire world. One of the first waves of migration from Africa was into India. Subsequent, more recent, waves of migration from other parts of the world have resulted in India being a genetic melting pot. Contemporary India has a rich tapestry of cultures and ecologies. There are about 400 tribal groups and more than 4000 groups of castes and subcastes, speaking dialects of 22 recognized languages belonging to four major language families. The contemporary social structure of Indian populations is characterized by endogamy with different degrees of porosity. The social structure, possibly coupled with large ecological heterogeneity, has resulted in considerable genetic diversity and local genetic differences within India. In this essay, we provide genetic evidence of how India may have been peopled, the nature and extent of its genetic diversity, and genetic structure among the extant populations of India.

OUR DNA IS A PALIMPSEST OF OUR HISTORY

Modern biology has provided powerful tools for reconstructing the history of the earth and its inhabitants, including humans. Central to this development has been the ability to study the genetic material of organisms, DNA. Scientists can extract DNA from microbes, plants, animals, and humans, including their fossilized remains. We can sequence DNA and extract valuable information from their sequences about the evolutionary past. Considerable human DNA sequence data have been obtained to date. These have been used to study our diversity and investigate how humans who reside in different geographical regions, or belong to distinct cultural groups, are genetically related. This is zoology and anthropology at the molecular level, and then the biologist turns into an historian.

The data generated by the human genome project indicate that two humans, selected randomly, are genetically ~99.9% identical in their

DNA sequences. Geneticists who study human genomic diversity therefore intensively focus their studies on the tiny fraction (~0.1%) of our genome in which we differ. This tiny fraction contains very rich information pertaining to our origins and diversity. Because the human genome comprises about three billion nucleotides, this tiny fraction (~0.1%) corresponds to about three million nucleotide differences.

Normally, only a fraction of the genetic variation of a large population of individuals is represented in a subset of those individuals. In other words, any subset of individuals of a larger set is genetically more homogeneous than the larger set. Residents of a restricted geographical region are usually descendants of a small ancestral group and, therefore, often have limited genetic variation among them. However, genetic variation between two ancestral groups is larger. Thus, descendants of a single ancestral group are genetically more similar than descendants from different ancestral groups. Therefore, by studying the patterns of variation in the genomes of contemporary individuals, resident in one or more geographical regions, it is possible to reconstruct their ancestral affiliations. The differences in DNA sequence in the tiny (0.1%) "variable" fraction of our genome also holds the key to why some individuals are susceptible to, whereas others are protected from, a specific disease. However, these disease-associated sequence differences, like those indicative of our historical origins, are not clustered, but spread across the entire genome.

Inferring human population history rests on a simple reality. New population groups arise or evolve from pre-existing groups. A population splits into subpopulations (population "fission") because of various cultural and demographic reasons and forces. In the past, when we were predominantly dependent on natural resources for our survival, increase of the numerical size of a population implied that there was pressure on natural resources. This pressure impinged on the survival of members of that group. Therefore, members of an expanding population would possibly have formed subgroups and moved away to new geographical locations in which natural resources were more abundant to form new subpopulations. Because these subgroups may have been small, each subgroup carried with them not the complete catalog, but only a fractional sample of the genomes present in the original population. This creates genomic diversity among the subpopulations. Furthermore, when subgroup sizes are small, demographic bottlenecks are created that would have exacerbated genomic diversity among the subpopulations. With the passage of time, subpopulations of an ancestral population diverge from one another. Despite these pressures and processes that enhance genomic differences among subpopulations of an ancestral population, some core genomic "signatures" are retained in the subpopulations. Thus, by studying the genome of individuals of contemporary subpopulations, it is possible to reconstruct their ancestries and ancestral relationships among the subpopulations.

After subpopulations emerge from an ancestral population, if they remain isolated, that is, if no mate-exchange or admixture takes place among them, they evolve independently. New mutations arise in each subpopulation. These mutations remain confined, because of lack of admixture, to the subpopulations in which they have arisen. Therefore, over time, genomic diversity is even more enhanced among the subpopulations.

Admixture, or the exchange of genes, allows a new genetic variation introduced by mutations to move from one subpopulation to another. Thus, admixture increases genetic similarities or affinities among subpopulations. The primary barriers to admixture are cultural differences, linguistic differences, and geographical distance. In general, unless there has been large-scale admixture on a continued basis, the longer two populations have been separated, the larger the genetic distance between them. Genetic distance, therefore, is a useful "clock" by which we can date evolutionary history. It must, however, be emphasized that various factors, especially natural selection, can, over time, increase or decrease the speed at which the clock ticks, thereby causing significant differences between the actual and estimated times of evolutionary events (Barreiro et al. 2008).

Anatomically, modern humans (*Homo sapiens sapiens*) evolved in Africa about 150,000 years ago and moved to other geographical regions between 125,000 and 60,000 years ago. There is now abundant evidence that human genomic diversity is greater in Africa than other global regions, indicating that Africa is likely to have been the source of subsequent human dispersals (Campbell and Tishkoff 2008).

Migration is not always symmetric with respect to gender; men appear to be more migratory than women. The genetic consequences of such gender asymmetry of migration can be revealed using genomic variations that are solely transmitted by either fathers or mothers, or transmitted to daughters or sons. Accounting for such gender differences is crucial because the dates of evolutionary events, such as migration to a new geographical area, are dependent on these differential rates. Geneticists have performed these analyses of maternal and paternal lineages using mitochondrial DNA (mtDNA), genetic material that is passed on by a mother to all of her children, therefore marking the genome of the last common female ancestor (female lineage), and, the Y chromosome, possessed only by males and transmitted by father to sons only, marking the genome of the last common male ancestor (male lineage).

The sequences contained in the mtDNA or the Y-chromosomal DNA change over time by mutation and accumulate as differences among individuals. The longer the elapsed time to their common ancestor, the larger the number of accumulated changes. Moreover, earlier changes are embedded in the DNA segment that carries later changes serving as a signature called a haplotype (haploid genotype). Haplotypes can be clustered into groups by the similarities of their DNA profiles called haplogroups and represent branches of maternal or paternal lineages. Consequently, maternal and paternal lineages can be defined by specific genetic variants (markers). Geneticists can estimate the age of a haplogroup; because a haplogroup is defined by the possession of a specific set of DNA variants, additional changes that appear can be used to estimate the age of the haplogroup. A haplogroup with a smaller number of accumulated

DNA changes is younger than another with a larger number of accumulated changes. If the rate of accumulation of change per year remains approximately constant, then one can estimate the time of accumulation from the observed number of changes. (Usually, such estimates are quite rough because of random fluctuations in accumulation of changes in different regions of the genome. Therefore, only a broad range of time can usually be ascribed.) The extant variation in our genomes thus carries footprints of the history of our species. Additionally, the pattern with which these mutations and polymorphisms (mutations that increase to high frequencies in a population) accumulate in our genomes is indicative of the dynamics of our population history (Nordborg 1997; Kingman 2000; Rosenberg and Nordborg 2002).

Humans who first migrated from Africa followed a "southern exit route" from the Horn of Africa across the mouth of the Red Sea along the coastline of India to southeastern Asia and Australia (Oppenheimer 2012). The major genetic evidence in support of this route is that the two major branches of the mtDNA L3 haplogroup (the haplogroup that is rooted in Africa), labeled as M and N haplogroups, outside Africa, have a very large number of local lineages in South Asia, and the antiquity of the N lineage found in Europe or the Near East is smaller than that in South Asia (Richards et al. 2006). The probable date of dispersal through the southern exit route was about 70,000–80,000 years ago (Fig. 1) (Lahr and Foley 1998). Archeological evidence in support of this route has been scanty, primarily because the coastlines of that period have become deeply submerged from the subsequent rapid increase in sea levels. There are strong indications, however, that a second and more recent human migration from out of Africa followed a "northern exit route" through the Nile Valley into Central Asia and then beyond, including into India.

PEOPLING OF INDIA, EARLY SETTLERS, AND CONTEMPORARY SOCIAL STRUCTURE

India has served as a major corridor for the migration of modern humans who started to dis-

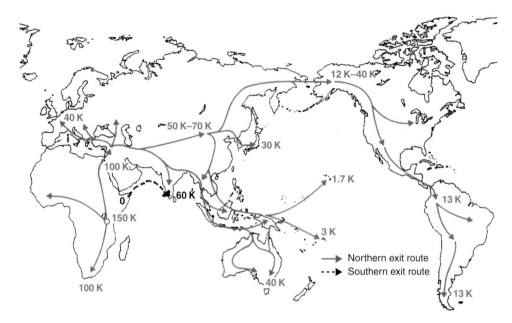

Figure 1. The great human exodus: The out-of-Africa journey and dispersal of anatomically modern humans. The numbers indicate the extimated dates in years before present.

perse out of Africa about 100,000 (perhaps, sometime between 125,000 and 60,000) years ago (Cann 2001). Nevertheless, the date of entry of modern humans into India remains uncertain. It is quite certain that by the middle of the Paleolithic period (50,000–20,000 years before present [ybp]) humans had spread to many parts of the subcontinent (Misra 1992, 2001). Also, modern human remains dating back to the late Pleistocene (55,000–25,000 ybp) have been found in India (Kennedy et al. 1987). Thus, India has been peopled by contemporary humans at least for the past 55,000 years (Fig. 1).

Molecular genetic evidence, for example, the pattern with which mutations have accumulated on the mtDNA in Indian populations, indicates that a major population expansion of modern humans took place within India (Majumder et al. 1999). Although the period of this demographic expansion remains uncertain, it has been speculated (Mountain et al. 1995) that it took place sometime during the range of 60,000–85,000 ybp. This expansion, followed by subsequent migration, appears to have resulted in the peopling of Southeast Asia and later (50,000–60,000 ybp) of Australia (Crow 1998).

An independent expansion of modern humans, ~60,000 ybp, appears to have taken place in southern China (Ballinger et al. 1992; Crow 1998), which may have resulted in human migration into India through the northeast corridor and also Southeast Asia.

Some recent archaeological finds from India indicate that the major route of dispersal to India from Africa was through the southern route (Mellars 2006). Molecular genetic studies, primarily those based on mtDNA and Y-chromosomal polymorphisms, have also favored a southern dispersal route (Underhill et al. 2001a,b; Forster 2004). The strongest genetic evidence in favor of an early southern exit into India comes from the observation that signatures possessed by Indian and other Asian populations are all derivatives of mitochondrial M and N haplogroups, which themselves derive from the L3 haplogroup now found only in Africa (Quintana-Murci et al. 1999). The southern exit hypothesis is also supported by analyses of mtDNA data from the Andaman Islands (Endicott et al. 2003; Kivisild et al. 2003; Thangaraj et al. 2005) and New Guinea (Forster et al. 2001). Furthermore, the Y-chromosomal haplogroups

C and D are found only in the Asian continent and Oceania (Endicott et al. 2003; Kivisild et al. 2003), but not in Eurasia or North Africa. Although available genomic data strongly indicate that India was peopled on one of the earliest waves of human migration from Africa and the dispersal took place via the southern exit route, a major limitation of more direct genetic evidence is the absence of reliable and comparable data from populations that lie on the southern exit route (Stringer 2000).

Contemporary India is a rich tapestry of largely intramarrying ethnic populations. These populations belong to a diverse set of cultures and language groups. There are four distinct language families in India, namely, Austro-Asiatic, Dravidian, Tibeto-Burman, and Indo-European, with a geographical distribution that is largely nonoverlapping within India. The Dravidian-speaking groups inhabit southern India, Indo-European speakers inhabit northern India, and Tibeto-Burman speakers are confined to northeastern India. In contrast, the numerically small group of Austro-Asiatic speakers, who are exclusively tribal, inhabit fragmented geographical areas of eastern and central India. Culturally, the vast majority of the people of India belong to either tribal or caste societies. The tribal populations are characterized by their traditional modes of subsistence: hunting and gathering, unorganized agriculture, slash and burn agriculture, and nomadism (practiced by a limited number of groups). They also have no written form of language and speak a variety of dialects. On the other hand, Hindu society in India (the numerically largest religious group) comprises castes that perform a wide range of occupations and have written forms of language. There is a long-standing debate about the genesis of the caste and tribal populations of India. One model suggests that the tribes and castes share considerable Pleistocene heritage with limited recent gene flow between them (Kivisild et al. 2003), whereas an opposite view concludes that caste and tribes have independent origins (Cordaux et al. 2004). Caste origins, however, appear to be complex. Origins of the same caste in different geographical regions appear to have been different, and genetic contributions to various castes

may have also been from different source populations (Basu et al. 2003; Majumder 2010).

There are more than 400 tribal groups and greater than 4000 groups of castes and subcastes in India. Although caste and tribe have been administrative social categories in British India, their existence as a social construct probably predates similar social categories found elsewhere in the world. Populations belonging to the caste fold have a ranked social order. There are four broad caste groups; however, commonly, the caste populations are now ranked as "high," "middle," and "low." Although the usage of the ranks as high, middle, and low can appear to have a value judgment attached to them, it is important to take cognizance of this extant hierarchy because, as Kosambi (1965) has pointed out, "stratification of Indian society reflects and explains a great deal of Indian history, if studied in the field without prejudice." Kosambi emphasized that the social and economic histories of ranked caste groups were different; the same also appears to be true of their genetic histories. There is virtually no evidence of the exchange of genes among tribal populations or between tribal and caste populations. There is also little exchange of genes among castes, primarily because of strict social rules of marriage within the caste system. Social stratification and norms governing mate-exchange among social strata impact on the genetic relationships of populations. Therefore, human geneticists have studied the genetic structures, similarities, and dissimilarities of ranked caste groups in India aiming to shed light on human history. Historical and anthropological studies suggest that, in the establishment of the caste system in India, there have been varying levels of admixture between the tribal people of India and the later immigrants who brought knowledge of agriculture, artisanship, and metallurgy from Central and West Asia. The migrants from Central and West Asia, who likely entered India through the northwestern corridor, spread to most areas of northern but not southern India. There is a distinct gradient of decreasing genetic similarity (representing a cline) of Indian populations with the West- and Central-Asian gene pools as we move eastward or southward from the northwestern corridor (Basu et al. 2003; Sengupta et al. 2006; Indian Genome Variation Consortium 2008; Reich et al. 2009). In other words, South and North India have had differential inputs of genes from Central and West Asia.

Who, then, are the earliest inhabitants of India? The Austro-Asiatic speakers are possibly the earliest settlers of India. They are exclusively tribal and show the highest frequencies of the ancient mtDNA lineage M. They also show the highest frequency (~20%) of the sublineage M2, which has the highest nucleotide variation within a fast evolving segment of the mitochondrial genome as compared with other sublineages. Recent results on Y-chromosomal markers provide further support for this inference. The Y lineage O-M95, found in high frequency in India, had originated in the Indian Austro-Asiatic populations around 65,000 ybp, very close to the estimated date of entry of the wave of human migration into India on the southern exit route from Africa. These findings are consistent with linguist Colin Renfrew's (Renfrew 2000) observation that the present distribution of the Austric language group is a result of the initial dispersal from Africa, whereas later agricultural dispersal can account for the distribution of the Elamo-Dravidian or Sino-Tibetan languages (the family to which Tibeto-Burman languages belong). However, recent studies have found that many Dravidian tribal populations also have M2 frequencies comparable to those of Austro-Asiatic tribal people (Kumar et al. 2008; Chandrasekhar et al. 2009). A recent (Chaubey et al. 2011) combined analysis of the uniparentally inherited Y-chromosomal markers with a large number of common single-nucleotide polymorphisms from the nuclear genome have, however, resulted in the proposal that the Austro-Asiatic speakers in India today are derived from dispersal from Southeast Asia, followed by extensive sex-specific admixture with local Indian populations.

Subjugation of an existing population by a relatively small group of highly organized, militarily powerful immigrants, the "elite dominance" model (Renfrew 2000), can obliterate preexisting genomic signatures through strong sexual exploitation by the immigrants, thereby presenting difficulties to our genomic inferences on the antiquities of populations. The antiquity of Dravidian speakers in India, who are not all tribal, but also belong to the organized caste system, has been enigmatic. Some historians have claimed that the Dravidians were widespread over nearly the entire landmass of India (Thapar 2004) and shared areas with Austro-Asiatic speakers. There is some genetic evidence (Basu et al. 2003) in support of this claim, and, furthermore, the Dravidian speakers probably retreated to their current habitat in southern India by the expansion of the more militarily powerful Indo-European speakers who arrived in India from Central Asia through the northwestern corridor. A contrary hypothesis, discussed later, has recently been postulated (Reich et al. 2009).

POPULATION DIVERSITY AND STRUCTURE: INFERENCES FROM mtDNA AND Y-CHROMOSOMAL DNA

Analysis of the genetic structure of Indian populations has shown that Indian ethnic populations, when grouped as tribal versus nontribal by geographical region of habitat or linguistic affiliation, have resulted from admixture of four or five ancestral populations (Indian Genome Variation Consortium 2008; Abdulla et al. 2009). One source of contribution is from Tibeto-Burman-speaking tribals who are distinct from the non-Tibeto-Burman speakers (Basu et al. 2003; Indian Genome Variation Consortium 2008; Abdulla et al. 2009). As ancestral sources of the Indian gene pool, in addition to the original African source population, West and Central Asia have been other major source contributors. However, these migrations from Asia may have taken place only in historical times, perhaps not earlier than 8000 ybp. Migration from West Asia was possibly associated with demic diffusion of agriculture, meaning the actual movement of people in the carriage of an idea or technology. The extent of genetic variation of female lineages (mtDNA) in India is rather restricted (Roychoudhury et al. 2001; Basu et al. 2003), indicating a small founding group of females. In contrast, the variation of male lineages (Y-chromosomal) is very high (Basu et al. 2003; Sengupta et al. 2006). This pattern may be indicative of sex-biased ancient gene flow into India with more male immigrants than female (Bamshad et al. 1998), possibly occurring within the last 5000 years through invasions and wars. This phenomenon obscures ancient genetic signatures and results in the quick introduction of high genetic variability, often mimicking extreme natural selection (Zerjal et al. 2003). The success of some of the Y-chromosomal haplotypes that arose in Central Asia to spread across vast regions of Eurasia (Zerjal et al. 2003), as well as South and Southeast Asia, is indicative of the "success" of the cultural and technological dominance of west Eurasia and Central Asia (Zerjal et al. 2003; Underhill et al. 2009).

The mtDNA lineage U, which is likely to have arisen in Central Asia, has a high frequency in India, implying that large-scale migration brought with it a large number of copies of this lineage into India. However, this lineage is composed of two deep sublineages, U2i and U2e, with an estimated split $\sim$50,000 years ago. The sublineage U2i is found in high frequency in India (particularly among tribes; $\sim$77%), but not in Europe ($\sim$0%), whereas U2e is found in high frequency in Europe ($>$10%), but not in India (it has very low frequencies among castes, but not among tribals). Thus, a substantial fraction of the U lineage, specifically, the U2i sublineage, may be indigenous to India (Ki-

visild et al. 1999; Basu et al. 2003; Sengupta et al. 2006). Analysis of the complete mtDNA genome sequence has revealed a large number of sequence variants within major haplogroups within Indian populations, many of which, however, are infrequent (Palanichamy et al. 2004). This indicates a common spread of the root haplotypes of haplogroups M, N, and R between 70,000 and 60,000 years ago along the southern exit route. This analysis has further revealed that entry of the haplogroup U2 postdates the earliest settlement along the southern route.

Central Asian populations are supposed to have been the major contributors to the Indian gene pool, particularly to the North Indian gene pool, and the migrants had supposedly moved into India through what is now Afghanistan and Pakistan. Using mtDNA variation data collated from various studies, we have previously shown (Basu et al. 2003) that populations of Central Asia and Pakistan show the lowest genetic distance with the North Indian populations ($F_{ST} = 0.017$) (see Box 1), higher distances ($F_{ST} = 0.042$) with the South Indian populations, and the highest values ($F_{ST} = 0.047$) with the northeast Indian populations; F_{ST} is a standard measure of genetic difference among populations derived from a common source population. Thus, northern Indian populations are genetically closer to Central Asians than populations of other geographical regions of India (Bamshad et al. 2001; Basu et al. 2003).

Although considerable cultural impact on social hierarchy and language in South Asia is

---

BOX 1. F_{ST}: FIXATION INDEX

F_{ST} is a measure of genetic differentiation among a number (k) of subpopulations. Consider a biallelic locus with alleles A and a whose frequencies in the ith subpopulation are, respectively, p_i and $q_i = 1 - p_i$ ($i = 1,2, \ldots, k$). Let p and q denote the frequencies of these two alleles, averaged over the k subpopulations. Then, the extent of genetic differentiation among the populations is measured by

$$F_{ST} = s^2/(pq),$$

in which $s^2 =$ variance of the frequency of allele A among the k subpopulations. When $k = 2$, F_{ST} can be used as a measure of genetic distance between two populations. $F_{ST} = 0$ indicates that the two populations are genetically identical, whereas $F_{ST} = 1$ indicates that they are as distinct as they can be.

---

attributable to the arrival of nomadic Central Asian pastoralists, studies using Y-chromosomal polymorphisms reveal that the influence of Central Asia on the preexisting gene pool was minor. Y-chromosomal data do not support models that invoke a pronounced recent genetic input from Central Asia to explain the observed genetic variation within South Asia. Genomic variation within Y-haplogroups R1a1 and R2 indicate demographic scenarios that are inconsistent with a recent single history. Deeper statistical analyses of the high-frequency R1a1 haplogroup chromosomes indicate independent recent histories of the Indus Valley and peninsular Indian region. These data are also more consistent with a peninsular origin of Dravidian speakers than a source with proximity to the Indus and significant genetic input resulting from demic diffusion associated with agriculture; rather, it indicates that pre-Holocene and Holocene-era, not Indo-European, expansions have shaped the distinctive South Asian Y-chromosome landscape (Sengupta et al. 2006).

NEW INFERENCES AND PARADIGMS FROM LARGE SETS OF GENOMIC MARKERS

Analyses of data on 405 single-nucleotide polymorphisms (SNPs) from 75 genes and a 5.2-Mb region on chromosome 22 in 1871 individuals from 55 diverse endogamous Indian populations (Indian Genome Variation Consortium 2008) have revealed that these populations form a genetic link bridging Caucasian and Asian populations. The HUGO Pan Asian SNP Consortium's study (Abdulla et al. 2009) also showed that most Indian populations share ancestry with European populations, which is consistent with recent observations and our understanding of the expansion of Indo-European-speaking populations. The study also provided strong evidence that the peopling of India (and Southeast Asia) was via a single primary wave of migration out of Africa (Abdulla et al. 2009). In the Indian Genome Variation Consortium (2008) study, genetic distances (F_{ST}) between pairs of populations were found to vary from 0.00 to 0.11, with a mean of 0.03, suggesting that the extent of overall differentiation was low.

Maximum F_{ST} values were observed among the tribal populations of different linguistic lineages. On a pan-India level, when populations were grouped by language or geographical region of habitat, the extent of genetic differentiation among linguistic or geographical groups was not statistically significant. However, grouping by ethnicity (caste and tribe) indicated significant differentiation, possibly caused by antiquity and isolation of the tribal compared with the caste populations. Although no clear geographical grouping of populations was found, ethnicity (tribal/nontribal) and language were the major determinants of genetic affinities among the populations of India.

Using more than 500,000 biallelic autosomal markers, Reich et al. (2009) have also found a north-to-south gradient of genetic proximity of Indian populations to western Eurasians/Central Asians. In general, the Central Asian populations were found to be genetically closer to the higher-ranking caste populations than to the middle- or lower-ranking caste populations. Among the higher-ranking caste populations, those of northern India are, however, genetically much closer to each other ($F_{ST} = 0.016$) than those of southern India ($F_{ST} = 0.031$). Phylogenetic analysis of Y-chromosomal data collated from various sources yields a similar picture.

Reich et al. (2009) have proposed that extant populations of India were "founded" by two hypothetical ancestral populations, one ancestral North Indian (ANI) and another ancestral South Indian (ASI). Presumably, these ancestral populations were derived from ancient humans who entered India via the southern and northern exit routes from out of Africa. All extant Indian populations are derived from admixture between these two putative ancestral populations, with the ANI contribution being higher among extant North Indian populations and that of ASI being higher among extant South Indian populations. In a more recent study (Moorjani et al. 2013), these investigators have shown that between 1900 and 4200 ybp, there was extensive admixture among Indian population groups, followed by a shift to endogamy. Because a large number of unbiasedly selected autosomal markers were used, this study does not suffer from the shortcomings of studies that used data from only one genetic locus (mitochondrial or Y-chromosomal). This model is simplistic, but intuitive and consistent with findings of earlier studies. It is simplistic because the origins of populations in the northeastern region of India cannot be explained by this model, and many past studies (cited earlier in this essay) have indicated genetic inputs into these populations from populations of Southeast Asia. More genome-wide studies with a larger sample of populations from India will provide further clarifications and insights into the population history of India.

ACKNOWLEDGMENTS

We are grateful to our collaborators and co-authors of papers, and also to those Indians who, over many years, have supported our genome diversity studies in many ways, including by donating blood samples. We are grateful to the Department of Biotechnology, Government of India, for providing financial support to our genome diversity studies in India.

REFERENCES

Abdulla MA, Ahmed I, Assawamakin A, Bhak J, Brahmachari SK, Calacal GC, Chaurasia A, Chen CH, Chen J, Chen YT, et al. 2009. Mapping human genetic diversity in Asia. *Science* **326:** 1541–1545.

Ballinger SW, Schurr TG, Torroni A, Gan YY, Hodge JA, Hassan K, Chen K-H, Wallace DC. 1992. Southeast Asian mitochondrial DNA analysis reveals continuity of ancient mongoloid migrations. *Genetics* **130:** 139–152.

Bamshad MJ, Watkins WS, Dixon ME, Jorde LB, Rao BB, Naidu JM, Prasad BV, Rasanayagam A, Hammer MF. 1998. Female gene flow stratifies Hindu castes. *Nature* **395:** 651–652.

Bamshad M, Kivisild T, Watkins WS, Dixon ME, Ricker CE, Rao BB, Naidu JM, Prasad BV, Reddy PG, Rasanayagam A, et al. 2001. Genetic evidence on the origins of Indian caste populations. *Genome Res* **11:** 994–1004.

Barreiro LB, Laval G, Quach H, Patin E, Quintana-Murci L. 2008. Natural selection has driven population differentiation in modern humans. *Nat Genet* **40:** 340–345.

Basu A, Mukherjee N, Roy S, Sengupta S, Banerjee S, Chakraborty M, Dey B, Roy M, Roy B, Bhattacharyya NP, et al. 2003. Ethnic India: A genomic view, with special reference to peopling and structure. *Genome Res* **13:** 2277–2290.

Campbell MC, Tishkoff SA. 2008. African genetic diversity: Implications for human demographic history, modern human origins, and complex disease mapping. *Annu Rev Genomics Hum Genet* **9:** 403–433.

Cann RL. 2001. Genetic clues to dispersal in human populations: Retracing the past from the present. *Science* **291:** 1742–1748.

Chandrasekar A, Kumar S, Sreenath J, Sarkar BN, Urade BP, Mallick S, Bandopadhyay SS, Barua P, Barik SS, Basu D, et al. 2009. Updating phylogeny of mitochondrial DNA macrohaplogroup m in India: Dispersal of modern human in South Asian corridor. *PLoS ONE* **4:** e7447.

Chaubey G, Metspalu M, Choi Y, Mägi R, Romero IG, Soares P, van Oven M, Behar DM, Rootsi S, Hudjashov G, et al. 2011. Population genetic structure in Indian Austroasiatic speakers: The role of landscape barriers and sex-specific admixture. *Mol Biol Evol* **28:** 1013–1024.

Cordaux R, Aunger R, Bentley G, Nasidze I, Sirajuddin SM, Stoneking M. 2004. Independent origins of Indian caste and tribal paternal lineages. *Curr Biol* **14:** 231–235.

Crow TJ. 1998. Was the speciation event on the Y chromosome? In *Abstracts of contributions to the dual congress*, p. 109. University of Witwatersrand Medical School, Johannesburg, South Africa.

Endicott P, Gilbert MT, Stringer C, Lalueza-Fox C, Willerslev E, Hansen AJ, Cooper A. 2003. The genetic origins of the Andaman Islanders. *Am J Hum Genet* **72:** 178–184.

Forster P. 2004. Ice ages and the mitochondrial DNA chronology of human dispersals: A review. *Philos Trans R Soc Lond B Biol Sci* **359:** 255–264; discussion 264.

Forster P, Torroni A, Renfrew C, Rohl A. 2001. Phylogenetic star contraction applied to Asian and Papuan mtDNA evolution. *Mol Biol Evol* **18:** 1864–1881.

Indian Genome Variation Consortium. 2008. Genetic landscape of the people of India: A canvas for disease gene exploration. *J Genet* **87:** 3–20.

Kennedy KAR, Deraniyagala SU, Roertgen WJ, Chiment J, Sisotell T. 1987. Upper Pleistocene fossil hominids from Sri Lanka. *Am J Phys Anthrop* **72:** 441–461.

Kingman JF. 2000. Origins of the coalescent: 1974–1982. *Genetics* **156:** 1461–1463.

Kivisild T, Bamshad MJ, Kaldma K, Metspalu M, Metspalu E, Reidla M, Laos S, Parik J, Watkins WS, Dixon ME, et al. 1999. Deep common ancestry of Indian and western-Eurasian mitochondrial DNA lineages. *Curr Biol* **9:** 1331–1334.

Kivisild T, Rootsi S, Metspalu M, Mastana S, Kaldma K, Parik J, Metspalu E, Adojaan M, Tolk HV, Stepanov V, et al. 2003. The genetic heritage of the earliest settlers persists both in Indian tribal and caste populations. *Am J Hum Genet* **72:** 313–332.

Kosambi DD. 1965. *The culture and civilisation of ancient India in historical outline*. Routledge & Kegan Paul, London.

Kumar S, Padmanabham PB, Ravuri RR, Uttaravalli K, Koneru P, Mukherjee PA, Das B, Kotal M, Xaviour D, Saheb SY, et al. 2008. The earliest settlers' antiquity and evolutionary history of Indian populations: Evidence from M2 mtDNA lineage. *BMC Evol Biol* **8:** 230.

Lahr MM, Foley RA. 1998. Towards a theory of modern human origins: Geography, demography, and diversity in recent human evolution. *Am J Phys Anthropol* **27:** 137–176.

Majumder PP. 2010. The human genetic history of South Asia. *Curr Biol* **20:** R184–R187.

Majumder PP, Roy B, Banerjee S, Chakraborty M, Dey B, Mukherjee N, Roy M, Thakurta PG, Sil SK. 1999. Human-specific insertion/deletion polymorphisms in Indian populations and their possible evolutionary implications. *Eur J Human Genet* **7:** 435–446.

Mellars P. 2006. Going east: New genetic and archaeological perspectives on the modern human colonization of Eurasia. *Science* **313:** 796–800.

Misra VN. 1992. Stone age in India: An ecological perspective. *Man Env* **14:** 17–64.

Misra VN. 2001. Prehistoric human colonization of India. *J Biosci* **26:** 491–531.

Moorjani P, Thangaraj K, Patterson N, Lipson M, Loh P-R, Govindaraj P, Berger B, Reich D, Singh L. 2013. Genetic evidence for recent population mixture in India. *Am J Hum Genet* **93:** 422–438.

Mountain JL, Hebert JM, Bhattacharyya S, Underhill PA, Ottolenghi C, Gadgil M, Cavalli-Sforza LL. 1995. Demographic history of India and mtDNA sequence diversity. *Am J Hum Genet* **56:** 979–992.

Nordborg M. 1997. Structured coalescent processes on different time scales. *Genetics* **146:** 1501–1514.

Oppenheimer S. 2012. Out-of-Africa, the peopling of continents and islands: Tracing uniparental gene trees across the map. *Phil Trans R Soc B* **367:** 770–784.

Palanichamy MG, Sun C, Agrawal S, Bandelt HJ, Kong QP, Khan F, Wang CY, Chaudhuri TK, Palla V, Zhang YP. 2004. Phylogeny of mitochondrial DNA macrohaplogroup N in India, based on complete sequencing: Implications for the peopling of South Asia. *Am J Hum Genet* **75:** 966–78.

Quintana-Murci L, Semino O, Bandelt HJ, Passarino G, McElreavey K, Santachiara-Benerecetti AS. 1999. Genetic evidence of an early exit of *Homo sapiens sapiens* from Africa through eastern Africa. *Nat Genet* **23:** 437–441.

Reich D, Thangaraj K, Patterson N, Price AL, Singh L. 2009. Reconstructing Indian population history. *Nature* **461:** 489–494.

Renfrew C. 2000. At the edge of knowability: Towards a prehistory of languages. *Cambridge Archaeol J* **10:** 7–34.

Richards M, Bandelt HJ, Kivisild T, Oppenheimer S. 2006. A model for the dispersal of modern humans out of Africa.

In *Human mitochondrial DNA and the evolution of Homo sapiens* (ed. Bandelt H-J, Macaulay V, Richards M), pp. 227–257. Springer, Berlin.

Rosenberg NA, Nordborg M. 2002. Genealogical trees, coalescent theory and the analysis of genetic polymorphisms. *Nat Rev Genet* **3:** 380–390.

Roychoudhury S, Roy S, Basu A, Banerjee R, Vishwanathan H, Usha Rani MV, Sil SK, Mitra M, Majumder PP. 2001. Genomic structures and population histories of linguistically distinct tribal groups of India. *Hum Genet* **109:** 339–350.

Sengupta S, Zhivotovsky LA, King R, Mehdi SQ, Edmonds CA, Chow CE, Lin AA, Mitra M, Sil SK, Ramesh A, et al. 2006. Polarity and temporality of high-resolution Y-chromosome distributions in India identify both indigenous and exogenous expansions and reveal minor genetic influence of Central Asian pastoralists. *Am J Hum Genet* **78:** 202–221.

Stringer C. 2000. Palaeoanthropology. Coasting out of Africa. *Nature* **405:** 24–25, 27.

Thangaraj K, Chaubey G, Kivisild T, Reddy AG, Singh VK, Rasalkar AA, Singh L. 2005. Reconstructing the origin of Andaman Islanders. *Science* **308:** 996.

Thapar R. 2004. *Early India: From the origins to AD 1300.* University of California Press, Oakland, CA.

Underhill PA, Passarino G, Lin AA, Marzuki S, Oefner PJ, Cavalli-Sforza LL, Chambers GK. 2001a. Maori origins, Y-chromosome haplotypes and implications for human history in the Pacific. *Hum Mutat* **17:** 271–280.

Underhill PA, Passarino G, Lin AA, Shen P, Mirazon LM, Foley RA, Oefner PJ, Cavalli-Sforza LL. 2001b. The phylogeography of Y chromosome binary haplotypes and the origins of modern human populations. *Ann Hum Genet* **65:** 43–62.

Underhill PA, Myres NM, Rootsi S, Metspalu M, Zhivotovsky LA, King RJ, Lin AA, Chow CE, Semino O, Battaglia V, et al. 2009. Separating the post-glacial coancestry of European and Asian Y chromosomes within haplogroup R1a. *Eur J Hum Genet* **18:** 479–484.

Zerjal T, Xue Y, Bertorelle G, Wells RS, Bao W, Zhu S, Qamar R, Ayub Q, Mohyuddin A, Fu S, et al. 2003. The genetic legacy of the Mongols. *Am J Hum Genet* **72:** 717–721.

How Genes Have Illuminated the History of Early Americans and Latino Americans

Andrés Ruiz-Linares

Department of Genetics, Evolution and Environment, University College London, London WC1E 6BT, United Kingdom

Correspondence: a.ruizlin@ucl.ac.uk

The American continent currently accounts for ~15% of the world population. Although first settled thousands of years ago and fitting its label as "the New World," the European colonial expansion initiated in the late 15th century resulted in people from virtually every corner of the globe subsequently settling in the Americas. The arrival of large numbers of immigrants led to a dramatic decline of the Native American population and extensive population mixing. A salient feature of the current human population of the Americas is, thus, its great diversity. The genetic variation of the Native peoples that recent immigrants encountered had been shaped by demographic events acting since the initial peopling of the continent. Similarly, but on a compressed timescale, the colonial history of the Americas has had a major impact on the genetic makeup of the current population of the continent. A range of genetic analyses has been used to study both the ancient settlement of the continent and more recent history of population mixing. Here, I show how these two strands of research overlap and make use of results from other scientific disciplines to produce a fuller picture of the settlement of the continent at different time periods. The biological diversity of the Americas also provides prominent examples of the complex interaction between biological and social factors in constructing human identities and of the difficulties in defining human populations.

A multiplicity of research approaches have been used to explore the original settlement of the American continent, often focusing on three prominent questions: (1) the route of entry of the initial settlers, (2) their time of arrival, and (3) the pattern of subsequent migration. These questions have been approached with variable degrees of success using various types of genetic markers examined in "Native" populations, defined on anthropological grounds (particularly language). Early studies used information from blood groups and proteins (Cavalli-Sforza et al. 1994) and were followed by DNA analyses mainly of mitochondrial DNA (mtDNA) (Forster et al. 1996; Tamm et al. 2007; Fagundes et al. 2008; Kitchen et al. 2008) and the Y chromosome (Lell et al. 1997; Bianchi et al. 1998; Karafet et al. 1999; Bortolini et al. 2003). The more recent studies have examined the human genome at increasing levels of resolution, from analyses with restricted sets of markers (Wang et al. 2007; Ray et al. 2010) to ongoing studies based on full genome sequences. Although a range of scenarios for the initial peopling of the Americas have been envisaged, genetic evidence points to the continent being settled by people migrating into the northwestern tip of the continent from Asia. This migration would have been facilitated by the existence, at that time, of a land bridge connecting Siberia to Alaska, which later was submerged beneath the Bering Strait by the rising sea level at the end of the last glaciation, around 15,000 years ago (Fiedel 2000). Genetic support for an American settlement from Eastern Siberia includes the finding that Native Americans are genetically most similar to North Asians (Cavalli-Sforza et al. 1994; Wang et al. 2007) and the existence of a gradient of declining genetic diversity from northwest North America southward (Wang et al. 2007; Reich et al. 2012). This gradient extends beyond that seen in the "Old World" for populations at increasing distance from Africa, possibly resulting from a sequence of population contractions that occurred as small groups of humans moved from settled areas into uninhabited territories (Ramachandran et al. 2005; Handley et al. 2007; Wang et al. 2007). The American continent, being the last major landmass to have been settled by humans, shows a low genetic diversity as compared with all other continents (Wang et al. 2007).

Estimating the date of the initial settlement of the Americas has proven a difficult and contentious issue. Geological information provides a key reference point in that because of extensive ice sheets covering North America at the peak of the last glaciation (around 20,000 years ago), the continent would have been impenetrable then (Fig. 1). Therefore, this leaves two broad opportunities for settlement: before or after this last glacial maximum (LGM). Calculating the time of initial settlement of the continent from genetic information requires a number of assumptions of which the exact validity is difficult to assess, including variation in factors such as population demography, mutation rates, and the influence of selection. Perhaps, not surprisingly, the range of genetic estimates for the time of human settlement of America is quite wide, extending to both sides of the glacial maximum. It is, however, encouraging that most of the recent estimates, based on increasingly larger

amounts of data and more sophisticated statistical methods, point to a settlement not long after the LGM. These estimates show greater consistency with the archaeological evidence, which, although itself not devoid of controversy, points to a human presence in the Americas by ∼14,000 years ago.

The pattern of migration into the continent has also been the subject of considerable disagreement. An influential model put forward in the mid-1980s posited that the settlement of the continent occurred in three sequential migratory waves from Asia, corresponding to the three major linguistic stocks in which the linguist Joseph Greenberg classified Native American languages (Greenberg et al. 1986; Greenberg 1987; Ruhlen 1991). The first migration would have given rise to a very large Amerind linguistic family comprising populations living all over the continent, whereas two subsequent migrations, restricted to North America and the Arctic, would be associated with populations speaking languages of the Na-Dene and Eskimo-Aleut linguistic families, respectively. Although early blood group and protein data were interpreted in support of the Greenberg model (Cavalli-Sforza et al. 1994), subsequent mtDNA and Y-chromosome analyses have been mostly interpreted as indicative of a single migration wave into the continent (Bonatto and Salzano 1997; Tamm et al. 2007; Fagundes et al. 2008; Kitchen et al. 2008). The recent genome-wide surveys of diversity with increasing resolution have, however, provided a different view. These are inconsistent with the single migration model and are more in line with the occurrence of multiple migrations (Fig. 1). Particularly strong support for several ancient migrations comes from a study based on a large survey of populations and using data for hundreds of thousands of genetic markers. With this type of data, it is possible to estimate the ancestry of every segment of DNA along the genome and state whether such a segment is of African, European, or Native American origin. Analyses can then focus only on the Native American segments of the genome (Fig. 2). This means that Native American individuals, and populations, that previously had to be excluded

Cite this article as *Cold Spring Harb Perspect Biol* doi: 10.1101/cshperspect.a008557

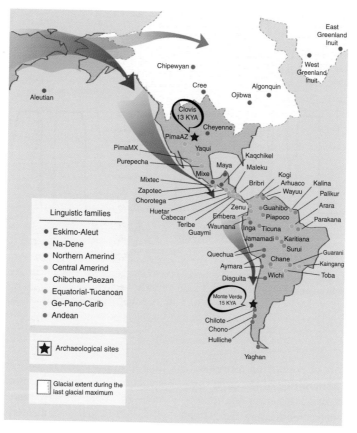

Figure 1. First peopling of the American continent. Settlement is thought to have occurred from Eastern Siberia through several waves of migration (arrows) across a land bridge connecting Siberia to Alaska, existing at the time. Crossing was impossible during the last glacial maximum (LGM) (~20,000 years ago) because of glaciers covering a large part of North America. Most genetic studies of contemporary Native Americans point to a settlement of the continents soon after the LGM, subsequent to the retreat of the ice sheets. Although classical studies associated initial settlement with the Clovis archaeological complex of North America (~13,000 years ago), older sites have been identified, including Monte Verde in South America (dated at ~15,000 years ago). The Native American populations placed on this map are those included in the phylogenetic tree shown in Figure 3. Analysis of genetic data from these populations is consistent with the important role of the coast during the initial settlement of the continent (Reich et al. 2012).

from study because of admixture with non-Natives can now be included, facilitating a more extensive population survey and reducing bias. These recent data have provided strong evidence that the Eskimo-Aleut, Na-Dene, and Amerind linguistic groups show evidence of differential gene flow from Asia, inconsistent with stemming from a discrete single colonization event with no subsequent migration (Fig. 1). Noticeably, although North American populations show evidence of multiple episodes of gene exchange with Asia, Native populations from

Mexico to the Southern tip of South America appear to stem from one colonization wave with no subsequent Asian gene flow. This observation agrees with the highly controversial proposal of grouping widely separated Native American languages into a single "Amerindian" linguistic family (Greenberg 1987; Ruhlen 1991). These data also confirm the correlation of population diversity with distance from the Bering Strait, in agreement with settlement in a north-to-south direction. Interestingly, this correlation increases when considering the

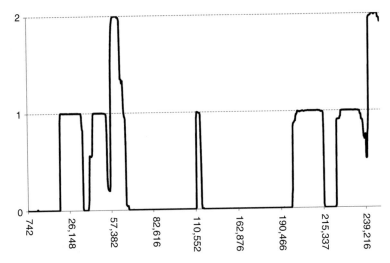

Figure 2. Inference of local ancestry along the two copies of chromosome 1 in an admixed Native American individual. The height of the thick line indicates local ancestry as the number of chromosome copies at that position that are estimated to be Native American (0, 1, or 2). Numbers on the x-axis refer to the position along chromosome 1 (in kilobases) of the genetic markers allowing inference of local ancestry. (Modified from data in Reich et al. 2012.)

coasts as facilitators of population movement, suggesting an important role of the coast during the initial population dispersals on the continent. A phylogenetic tree relating the Native American populations examined in that survey is also consistent with the north-to-south settlement of the continent, as it shows a sequence of major population splits separating groups of populations mostly along a north-to-south axis (Figs. 1 and 3). Consistent with some degree of parallel evolution for languages and genes (Cavalli-Sforza et al. 1994), resulting from population separation followed by relative isolation, the major clusters of populations in this genetic tree show a broad correspondence with the linguistic affiliation of the populations (Fig. 3).

The recent study by Reich et al. (2012) illustrates the potential of high-density genotyping for extending studies focused on the original settlement of the Americas to Native individuals with evidence of admixture with recent immigrants. This admixture is extensive across the continent and involves not only Natives but also the general population, particularly in the countries of what is now referred to as Latin America. Historically, a major driver behind population mixing in this region was the fact

that immigrants from Spain and Portugal, particularly in the early phases of the colonial expansion, were mostly men (Boyd-Bowman 1973). It is well documented that many Conquistadors had children with Native women, the most famous example possibly being that of the Conquistador of Mexico, Hernán Cortez, and the Nahua woman known as "Malinche" (Fig. 4). This "sex-biased" pattern of mating between immigrant men and Native women had been alluded to by historians (Morner 1967), but it was only with mtDNA and Y-chromosome studies that the full genetic impact of this feature became apparent. Because mtDNA and the Y chromosome are only transmitted by mothers and fathers, respectively, they allow the inference of the maternal (mtDNA) and paternal (Y-chromosome) ancestry of individuals. One of the first such studies was performed in Antioquia (Colombia), a population traditionally considered as mainly of Spanish descent. Consistent with this view, it was found that >90% of men in Antioquia had Y-chromosome lineages of European origin (Carvajal-Carmona et al. 2000). Surprisingly, when examining their mtDNA, a sharply different picture was observed. In 90% of individuals, maternal

Cite this article as *Cold Spring Harb Perspect Biol* doi: 10.1101/cshperspect.a008557

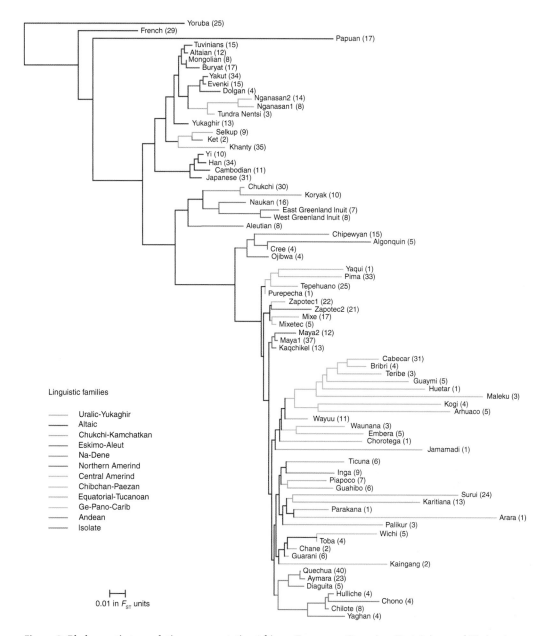

Figure 3. Phylogenetic tree relating representative African, European, Oceanian, East Asian, and Native American populations based on high-density genetic marker data. For this analysis, Native American data used was restricted to genome segments of confirmed Native ancestry (determined as in Fig. 2). The tree branches are color-coded to represent the linguistic affiliation of the populations, as shown in the inset. Numbers in parentheses refer to sample size in each population. The length of the branches on this tree is proportional to a measure of genetic differentiation (F_{ST}). (From Reich et al. 2012; reprinted, with permission, from the author.)

Figure 4. The Native American woman known as "Malinche" or Malintzin (her Nahuatl name) was the interpreter and mistress of the Spanish Conquistador, Hernán Cortés. In 1523, she gave him a son, Martín, who is one of the first recorded individuals of mixed Native–European ancestry born in the Americas. Such offspring between immigrant men and Native women were a common occurrence in early colonial Latin America. The drawing shown is from the late 16th century "Codex Tlaxcala" and represents a meeting between the Mexican ruler, Moctezuma, and Hernán Cortés, with Malintzin (on the *right*) translating.

ancestry was Native American (Carvajal-Carmona et al. 2000). Similar observations have now been made in many Latin American countries (Alves-Silva et al. 2000; Green et al. 2000; Carvalho-Silva et al. 2001; Marrero et al. 2007), although with a considerable variation in ancestry proportions between them (Fig. 5). These studies, in addition, show a higher African ancestry with mtDNA than the Y chromosome, indicating that, historically, admixture with Africans has also mostly involved African women.

The large variation in ancestry seen across Latin America relates to differences in pre-Columbian Native population density and the pattern of recent immigration into specific regions of the continent. For instance, most studies performed so far have mainly focused on areas with little documented African immigration and consistently show a relatively low African

genetic ancestry. In these population samples, the variation in individual European and Native American ancestry is very large, to the extent that it overlaps with that seen in Native population samples (Fig. 6). The variation in individual ancestry seen in these samples thus effaces their designation as "Native" or "non-Native." This observation punctuates the interest of incorporating admixed Latin American populations, traditionally considered non-Native, into studies on the initial settlement of the continent. Similar to what has been performed in the recent survey by Reich et al. (2012), the inference of ancestry of each genome segment in Latin Americans could be used to focus solely on Native American segments of the genome. This is an avenue of research that is just beginning to be explored and shows great potential for the future. It promises to be of particular importance for the analysis of regions where anthropologically recognizable Native populations and individuals are virtually nonexistent, as they have been absorbed into the current mixed population. This is the case for the many areas that were relatively sparsely populated in pre-Columbian times and, subsequently, received a large flow of immigrants, such as from the Caribbean and many parts of North and South America. Consequently, estimation of individual ancestry along the genome will facilitate denser demographic history analyses across the Americas, as well as a reexamination of the original settlement of the continent based on a more comprehensive population sampling.

Other than being informative for addressing questions of population history, the study of Latin American populations promises to facilitate the genetic characterization of biological attributes differentiated among the populations that participated in admixture on the continent. For instance, a range of facial features differ between Native Americans and Europeans, and the genetic study of admixed Latin Americans promises to help in the identification of genes explaining variation in facial appearance. Such research is of interest for understanding disorders of craniofacial development and could also have forensic applications. Another example is type 2 diabetes (T2D), a dis-

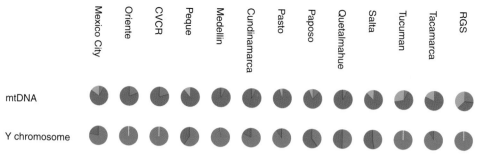

Figure 5. Proportion of Native American, European, and African ancestry in 13 Latin American populations estimated using mtDNA and Y-chromosome markers. Samples from urban centers in five countries were examined (Mexico—Mexico City; Guatemala—Oriente; Costa Rica—Central Valley of Costa Rica [CVCR]; Colombia—Peque, Medellín, and Cundinamarca; Chile—Paposo and Quetalmahue; Argentina—Salta, Tucuman, and Tacamarca; and Brazil—Rio Grande do Sul [RGS]). Ancestry proportion (fraction of the pie chart) is color-coded: African (green), European (blue), and Native American (red). (Data for 20 individuals per population are from Wang et al. 2007, Yang et al. 2010, and NN Yang et al. unpubl.)

ease that has a very high frequency in Native Americans and for which a higher risk is associated with increased Native American ancestry. This observation led to the proposal of the "thrifty genotype" hypothesis, which posits that the increased risk of T2D in Native Americans results from genetic adaption to a low-calorie/high-exercise way of life that became detrimental with the recent change to a high-calorie/

low-exercise lifestyle (Pollard 2008). The study of large, carefully characterized samples from Latin American populations offers a unique opportunity for conducting a detailed assessment of this hypothesis. The identification of genes explaining the variable frequency of diseases between populations (such as T2D) will be an important step forward in the development of novel, more effective (even individualized) dis-

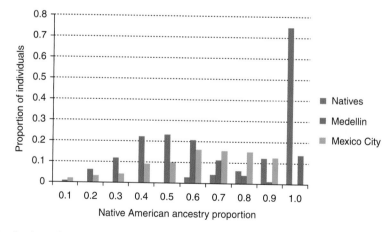

Figure 6. Distribution of individual Native ancestry estimated in samples from Native Americans (shown in blue) and two Latin American urban centers (850 residents of Medellín, Colombia shown in red, and 220 residents of Mexico City, Mexico shown in green). For each of the three population samples, the x-axis indicates the proportion of Native ancestry (in 0.1 unit intervals) and the y-axis indicates the proportion of individuals with that ancestry estimate. The Native American group includes 225 individuals from North, Central, and South America. Ancestry proportions were estimated using autosomal genetic markers. (Data from Florez et al. 2009, Campbell et al. 2011, and Reich et al. 2012.)

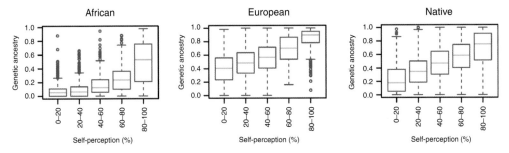

Figure 7. Box plots displaying the relationship of individual genetic ancestry estimates to self-perceived ancestry in 7342 Latin Americans (from Mexico, Colombia, Peru, Chile, and Brazil). Self-perception was categorized into 20% bands for African, European, and Native American ancestry. There is a highly significant correlation between the genetic estimate and self-perception for each continental ancestry component. However, there is a trend at higher Native American and African ancestry for self-perception to exceed the genetic estimates. Correspondingly, at lower European ancestry, there is a trend for the genetic estimates to exceed self-perception. Further analyses show that pigmentation impacts on these differences. Individuals with lower skin pigmentation tend to overestimate their European ancestry, whereas individuals with higher pigmentation overestimate their Native American and African ancestries. Orange lines indicate the median and the blue boxes are delimited by the 25th and 75th percentiles. (From AR Ruiz-Linares et al., in press; with permission from the author.)

ease-management strategies that account for human population diversity.

The overlap of individual genetic ancestry estimates, seen in Latin and Native American populations (Fig. 5), raises the question of the relationship of these estimates to the perception that individuals have of their own ancestry. A recent analysis of a large sample of individuals from five Latin American countries found a highly significant correlation between self-perception and genetically estimated ancestry (Fig. 7). However, this study also found evidence that self-perception is biased. A particularly clear bias involves pigmentation: individuals with greater pigmentation tend to overestimate their Native and African ancestry, whereas individuals with lighter pigmentation tend to overestimate their European ancestry (AR Ruiz-Linares, in press). Statistically significant differences were also observed between countries, pointing to the influence of social factors in self-perception. Consistent with this observation, social scientists have argued that, in Latin America, self-identification as Native or non-Native is often strongly influenced by social cues (Wade 2010).

The insights into the initial settlement of the continent provided by the genetic study of Native Americans illustrate the fact that a popula-

tion sampling that maximizes diversity based on anthropological grounds (such as language) can facilitate investigations concerned with the first settlement of the continent. However, the analysis of intercontinental admixture both in Native and non-Native Latin American populations show some of the complexities of defining human groups, with population labels suggesting a potentially misleading genetic singularity. The biological reality is that of a gradient in the genetic makeup of these populations (and individuals) involving various degrees of mixture between the initial settlers of the continent and more recent immigrants. The genetic diversity of Latin Americans is, thus, a prominent example of the fuzzy meaning of the labels used to refer to human populations. Although these labels can assist in study design and facilitate certain historical inferences, ethnicity, race, and other such terms are social constructs devoid of a clear-cut biological meaning.

REFERENCES

Alves-Silva J, da Silva Santos M, Guimarães PE, Ferreira AC, Bandelt HJ, Pena SD, Prado VF. 2000. The ancestry of Brazilian mtDNA lineages. *Am J Hum Genet* **67:** 444–461.

Bianchi NO, Catanesi CI, Bailliet G, Martinez-Marignac VL, Bravi CM, Vidal-Rioja LB, Herrera RJ, López-Camelo JS.

 Cite this article as *Cold Spring Harb Perspect Biol* doi: 10.1101/cshperspect.a008557

1998. Characterization of ancestral and derived Y-chromosome haplotypes of New World native populations. *Am J Hum Genet* **63:** 1862–1871.

Bonatto SL, Salzano FM. 1997. A single and early migration for the peopling of the Americas supported by mitochondrial DNA sequence data. *Proc Natl Acad Sci* **94:** 1866–1871.

Bortolini MC, Salzano FM, Thomas MG, Stuart S, Nasanen SP, Bau CH, Hutz MH, Layrisse Z, Petzl-Erler ML, Tsuneto LT, et al. 2003. Y-chromosome evidence for differing ancient demographic histories in the Americas. *Am J Hum Genet* **73:** 524–539.

Boyd-Bowman P. 1973. Patterns of Spanish emigration to the New World (1493–1580). Council on International Studies, State University of New York at Buffalo, Buffalo, NY.

Carvajal-Carmona LG, Soto ID, Pineda N, Ortíz-Barrientos D, Duque C, Ospina-Duque J, McCarthy M, Montoya P, Alvarez VM, Bedoya G, et al. 2000. Strong Amerind/white sex bias and a possible Sephardic contribution among the founders of a population in northwest Colombia. *Am J Hum Genet* **675:** 1287–1295.

Carvalho-Silva DR, Santos FR, Rocha J, Pena SD. 2001. The phylogeography of Brazilian Y-chromosome lineages 1. *Am J Hum Genet* **68:** 281–286.

Cavalli-Sforza LL, Menozzi P, Piazza A. 1994. *The history and geography of human genes.* Princeton University Press, Princeton, NJ.

Fagundes NJ, Kanitz R, Eckert R, Valls AC, Bogo MR, Salzano FM, Smith DG, Silva WA Jr, Zago MA, Ribeiro-dos-Santos AK, et al. 2008. Mitochondrial population genomics supports a single pre-Clovis origin with a coastal route for the peopling of the Americas. *Am J Hum Genet* **82:** 583–592.

Fiedel SJ. 2000. The peopling of the New World: Present evidence, new theories, and future directions. *J Archaeol Res* **8:** 39–103.

Forster P, Harding R, Torroni A, Bandelt HJ. 1996. Origin and evolution of Native American mtDNA variation: A reappraisal. *Am J Hum Genet* **59:** 935–945.

Green LD, Derr JN, Knight A. 2000. mtDNA affinities of the peoples of North-Central Mexico. *Am J Hum Genet* **66:** 989–998.

Greenberg JH. 1987. *Language in the Americas.* Stanford University Press, Stanford, CA.

Greenberg JH, Turner CG II, Zegura SL. 1986. The settlement of the Americas: A comparison of the linguistic, dental, and genetic evidence. *Curr Anthropol* **27:** 477–497.

Handley LJ, Manica A, Goudet J, Balloux F. 2007. Going the distance: Human population genetics in a clinal world. *Trends Genet* **23:** 432–439.

Karafet TM, Zegura SL, Posukh O, Osipova L, Bergen A, Long J, Goldman D, Klitz W, Harihara S, de Knijff P, et al. 1999. Ancestral Asian source(s) of new world Y-chromosome founder haplotypes. *Am J Hum Genet* **64:** 817–831.

Kitchen A, Miyamoto MM, Mulligan CJ. 2008. A three-stage colonization model for the peopling of the Americas. *PLoS ONE* **3:** e1596.

Lell JT, Brown MD, Schurr TG, Sukernik RI, Starikovskaya YB, Torroni A, Moore LG, Troup GM, Wallace DC. 1997. Y chromosome polymorphisms in Native American and Siberian populations: Identification of Native American Y chromosome haplotypes. *Hum Genet* **100:** 536–543.

Marrero AR, Bravi C, Stuart S, Long JC, Pereira das Neves Leite F, Kommers T, Carvalho CM, Pena SD, Ruiz-Linares A, Salzano FM, et al. 2007. Pre- and post-Columbian gene and cultural continuity: The case of the Gaucho from southern Brazil. *Hum Hered* **64:** 160–171.

Morner M. 1967. *Race mixture in the history of Latin America.* Little Brown, New York.

Pollard TM. 2008. *Western diseases: An evolutionary perspective.* Cambridge University Press, New York.

Ramachandran S, Deshpande O, Roseman CC, Rosenberg NA, Feldman MW, Cavalli-Sforza LL. 2005. Support from the relationship of genetic and geographic distance in human populations for a serial founder effect originating in Africa. *Proc Natl Acad Sci* **102:** 15942–15947.

Ray N, Wegmann D, Fagundes NJ, Wang S, Ruiz-Linares A, Excoffier L. 2010. A statistical evaluation of models for the initial settlement of the American continent emphasizes the importance of gene flow with Asia. *Mol Biol Evol* **27:** 337–345.

Reich D, Patterson N, Campbell D, Tandon A, Mazieres S, Ray N, Parra MV, Rojas W, Duque C, Mesa N, et al. 2012. Reconstructing Native American population history. *Nature* **488:** 370–374.

Ruhlen M. 1991. *A guide to the world's languages.* Stanford University Press, Stanford, CA.

Tamm E, Kivisild T, Reidla M, Metspalu M, Smith DG, Mulligan CJ, Bravi CM, Rickards O, Martinez-Labarga C, Khusnutdinova EK, et al. 2007. Beringian standstill and spread of Native American founders. *PLoS ONE* **2:** 1–6.

Wade P. 2010. *Race and ethnicity in Latin America.* Pluto, London.

Wang S, Lewis CM, Jakobsson M, Ramachandran S, Ray N, Bedoya G, Rojas W, Parra MV, Molina JA, Gallo C, et al. 2007. Genetic variation and population structure in native Americans. *PLoS Genet* **3:** e185.

Can Genetics Help Us Understand Indian Social History?

Romila Thapar

Centre for Historical Studies, Jawaharlal Nehru University, New Delhi 110067, India

Correspondence: romila.thapar@gmail.com

Attempts have been made recently to determine the identity of the so-called "Aryans" as components of the Indian population by using DNA analysis. This is largely to ascertain whether they were indigenous to India or were foreign arrivals. Similar attempts have been made to trace the origins of caste groups on the basis of varna identities and record their distribution. The results so far have been contradictory and, therefore, not of much help to social historians. There are problems in the defining of categories and the techniques of analysis. Aryan is a linguistic and cultural category and not a biological one. Caste groups have no well-defined and invariable boundaries despite marriage codes. Various other categories have been assimilated into particular castes as part of the evolution of social history on the subcontinent. A few examples of these are discussed. The problems with using DNA analysis are also touched on.

The substantive part of the text deals with how contemporary historians have examined the evidence on race and caste in Indian history. They have discarded the first and have found the earlier definition of the second to be based on theory without adequate attention to caste as it actually functioned. Caste identity, particularly among the higher castes, and always assumed to be controlled by strict rules of marriage codes, has, in fact, been open to the assimilation of diverse groups for historical reasons. None of these identities were free of biological admixture.

Viewing the human populations of India since early history, it is assumed that genetic analyses can provide some evidence of their origins and of their social history. But the categories used in these analyses have emerged out of historical conditioning and, consequently, tend to bring the argument back to the historical. Current classifications of Indians by caste are almost entirely socially determined, barring the single exception of the biological difference of gender. The commonly used categories are race, caste, language, region, and some isolated habitats. Of these, race has been a dominant category in colonial scholarship during the last two centuries and caste goes back to much earlier times. Language was part of a larger differentiation. Region and habitat are of recent vintage.

Race as a category is unknown to traditional Indian classifications. It was imported from Britain and was used mainly by Europeans writing on Indian society, although subsequently it came to be used by Indians as well. What we call caste is referred to in Sanskrit sources by two concepts—varna and jati. They both assume identity through being born into a group and following a particular occupation. The first literally means color and was used rather metaphorically as such when referring to the four major caste groups—Brahmana, Kshatriya, Vaishya, and Shudra, sometimes listed as white, red, yellow, and black. Sociologists sometimes render the concept of varna as ritual or status ranking or alternately as the broader divisions of society. In the latter case, the meaning can be traced to "cover." The word jati comes from the root *ja*, meaning to be born. This refers directly to identity based on birth and also extends to occupation, but it is a narrower category than varna and there are hundreds of jatis observing various rules of endogamy and exogamy.

The Brahmanical normative texts, such as the *Dharmashastras* that incorporate the social codes, projected a society that closely observed the rules and allowed for little flexibility. This may be one reason why it was thought that caste society would lend itself to genetic analyses. However, the work of social historians has shown that this was not a description of how society actually functioned because a range of other texts provide evidence that the codes may have been formally observed but in fact were being broken. In effect, the caste would conform to its rank in the hierarchy of castes but the group that constituted the caste could change.

The beginnings of Indian history have been beset with contradictory theories about the origins of the peoples that created the fundamental cultures of India. These contradictions focus on the question of the origin of "the Aryans" and subsequently of "the Dravidians" (Dolgin 2009). Were the Aryans a people from Central Asia with cultural similarities to those of west Eurasia who migrated and settled in northwest India from where they spread across the subcontinent? Or were the Aryans indigenous to India, and according to some, spread their culture from out of India westward? Attempts have been made to use genetic analyses to determine the identity of various population groups and attempt an answer to these questions. The prob-

lem remains, however, because no such groups have survived as a distinct entity from that period. Furthermore, the categories have been and constantly are redefined by historians and social scientists, and the redefinitions no longer allow an unimpaired existence across the centuries.

Historical evidence about the populations of that period—the second millennium BC—is dependent on archaeology and history. Attempts have been made to corelate archaeological cultures in Central Asia with textual references from the *Avesta* composed in Old Iranian, and the *Rigveda* composed in Indo-Aryan. The closest is the Bactro-Margiana Archaeological Complex (BMAC), sometimes called the Oxus Civilization, dating to 2300–1900 BC, but even this is problematic (Staal 2008). Skeletal material from archaeological sites cannot be identified as "Aryan" because this is not a biological concept.

The western Orientalist reconstruction of Indian history begins with the Aryans as founders of Indian civilization. This, of course, is not the narrative in the early Indian Puranic "histories," which make no mention of any Aryans in this role. The invention of Aryan foundations, therefore, is a 19th century way of reading the beginnings of Indian history. The Vedic texts earlier were revered as texts of religious belief and practice and not as recording the beginnings of Indian history.

The word "arya" has two social and cultural connotations widely used in history but separated by 3000 years. The first is its meaning from Indian texts (Thapar 2008), but this meaning undergoes change through the centuries. Initially, it is a rather vague word used to describe those thought to be socially and culturally acceptable and worthy of respect, who used Indo-Aryan/Vedic Sanskrit as their language, and were relatively well-to-do. In the *Rigveda*, the Aryan is differentiated from the dasa, a word used to collectively describe the other component of the population. The dasa is associated with alien culture and speech, unfamiliar rituals, and with evil and darkness. Some, however, were rich in cattle wealth and subjected to raids by the aryas. Presumably it was these rich dasas who, apart from the aryas, occasionally were the

Cite this article as *Cold Spring Harb Perspect Biol* doi: 10.1101/cshperspect.a008599

patrons of Brahmin rituals. Gradually, the use of the word was extended to mean a person to be respected and was therefore used for Buddhist monks and for royalty. By the turn of the Christian era, it came to be used for the dvija/twice-born upper castes, whereas the lower castes are described as non-aryas (Manu 1991).

This was read by 19th-century western Orientalists as referring to two races—the Aryan and the Dravidian—represented by the dasas. Both were language labels that came to be used interchangeably with race. The correct usage is Aryan-speaking and Dravidian-speaking peoples, but Aryan and Dravidian used as short cuts took on the meaning of race, ethnicity, and culture, with the imprint of race being the stronger. Similarities between Indo-Aryan and early European languages provided a basis for maintaining that there was an original Indo-European language spoken in Central Asia and that the speakers divided into two, with one branch going westward and the other finally arriving in India. The language and culture of the latter became the foundation of Hindu culture. The Aryans inevitably were the progenitors of the upper castes and the dasas, said to have been subordinated by the Aryans, of the lower castes (Trautmann 1997; Thapar 2008).

The construction of the terms Aryan and Dravidian grew from the study of Sanskrit and Old Tamil. Sanskrit was recognized as having cognates with Persian and Greek, and a theory of linguistic monogenesis was put forward by Sir William Jones at the end of the 18th century. A few decades later, Ellis and Caldwell argued for a parallel Dravidian group of languages. The Austro-Asiatic group, chiefly Munda, was regarded as dissimilar to both the others. The picture was one of multilingualism with each language being equated with a race. The equation was strengthened by the theories of what came to be called "race science" in Europe, which maintained that identification by race was based on scientific truth. Classification according to race was influenced by biological studies of the time and some degree of social Darwinism. Colonialism appropriated the notion that the colonizers were superior and the colonized, inferior.

The 19th century study of philology in Europe, which advanced rapidly after the inclusion of Sanskrit as a comparative language, made a particular impression on German Romanticism (Schwab 1984). Sanskrit was now spoken of as the ancestral language, a theory also advanced by those theosophists who had made India their home. This led to a search for an Indo-European homeland in Central Asia, later to be regarded as the source for proto-Indo-European speech. Many 19th century Europeans were searching for the "pure" Aryans and among them was the Comte de Gobineau.

This in part triggered the change to look for the homeland in Europe and the notion that the original Aryan may have been a blonde Nordic (Poliakov 1974). Measurements of the nasal and cephalic indices became crucial to proving racial purity. This turned into a major exercise in India where there was a project to confirm the classification of races by these measurements (Risley and Crooke 1915).

Max Mueller's study of the Vedas in the latter part of the 19th century and his reconstruction of the early Indian past put the stamp of authority on the theory of an Aryan race. Although well aware of not confusing race and language, he nevertheless proceeded to do just that, as did other scholars of the time. He argued that the word varna used for caste in the Vedic texts meaning color referred to the fair-skinned Aryans and the dark-skinned dasas. Caste was much emphasized as the distinction between the upper caste Aryans and the lower caste Dravidians. One of the epithets used for dasas was *a-nas*, which he read as without a nose, and this was at a time when the measuring of nasal indices was regarded as firm evidence, although the alternative reading is *an-as*, without a mouth (i.e., not speaking a comprehensible language). Situations of conflicts between the two are mentioned and this was taken as proof that the Aryans invaded northwestern India and established themselves as conquerors in the mid-second millennium BC. This idea has now been discarded and the preference is for a graduated migration and much mixing with existing inhabitants—except among those few who continue to insist that the Aryans were indigenous. Differences in language and rituals were obvious. The structure of Dravidian languages was not the same as that of Indo-Aryan and therefore the racial distinction was also underlined (Ramaswamy 1997; Trautmann 2006).

The Indian reaction to these theories at the end of the 19th century was an acceptance of their main ideas, which suited the identities sought by the emerging middle class. But these ideas were reformulated and eventually came to be used politically. The category of race was gradually replaced by caste. Because there was no word for race in Sanskrit, the term jati, which together with varna was used for caste, came to be used for race. The preference for jati was because its root came from ja, birth. This was partly responsible for the change.

Caste as a category of exclusion or as a part of social stratification has been a way of ordering societal characteristics of the Indian subcontinent for the past 2000 years. The stratification is based on dividing society into privileged and nonprivileged groups, the first constituted by the upper castes and the second by the lower castes. The normative condition was that social mobility was not to be allowed into the former, whereas control over the latter was not in effect possible. Three broad groups were outside the caste structure and often referred to as "mleccha": the untouchables/Dalits regarded as ritually polluted and who were physically segregated and formed a separate social system of their own, the tribal societies of the forest habitats, and those that came from other lands and cultures. There was a belief that an immobile, frozen society could be created and could function without change through enforcing the code of caste functioning, especially rules of marriage and inheritance.

The two systems of caste organization were juxtaposed. One was varna, where society was divided into four hierarchical components or castes. The normative texts, such as the *Dharmashastra* of Manu, were the social codes that in theory determined identity and functions. The relationship of one to the other is crucial to the system and conformation at least in theory, necessary. The other associated system was jati, in which there was again a hierarchical division.

Here the rules of birth and marriage were suggestive of the functioning of clan societies but with an emphasis on occupation, and written codes were largely absent. There was some attempt to find equivalences with the varna categories but this was problematic as jati hierarchies prevailed largely in the lower two varnas and as such they were more flexible. Neither system was as rigid as was hoped for in the normative texts. The mixing of castes was regarded as social degradation, yet many castes, high and low, resulted from such mixing. Among the more influential of these was the Kayastha caste of scribes and administrators. The exclusivity of caste was maintained but entry into an upper caste status could not be barred to other castes as is shown by the contradictions in the texts. Some royal families were of obscure origin but the genealogies made for them linked them to high castes, a case in point being the 18th century Maratha ruler, Shivaji. However, both the Orientalists—brought up on a diet of normative texts—and the Indian middle class, drawing inspiration from the former, believed in the purity of descent of each varna.

Indian interpretations of the theory of the Aryan race went from one extreme to another and shifted the identity of Aryan. Jyotiba Phule in Maharashtra turned the theory upside down, as it were (Deshpande 2002). He interpreted it as indicating that the original inhabitants were the lower castes who were made servile by an invasion of Brahmin/Aryan aliens. The latter took away the land of the former through guile. He drew on various myths as support for his ideas. The cultures that existed before the coming of the Aryans were therefore the creation of the lower castes. This argument has been useful to the identity of lower castes, particularly after the discovery of the Indus Civilization in the early 20th century, with which civilization they claim close links. Phule's reading shifted the focus from race onto caste with caste as the differentiating feature.

Upper caste authors ignored Phule. Thus, B.G. Tilak, for example, argued that the Aryans trekked from the Arctic where they originated and one branch came to India. Later, when there was an insistence that the Aryans were indige-

Cite this article as *Cold Spring Harb Perspect Biol* doi: 10.1101/cshperspect.a008599

nous to India, Tilak's theory continued to be accepted, but the location of the North Pole had to be shifted to within British Indian territory (Das 1920)!

The reverse of Phule's theory was propounded by Dayanand Sarasvati the founder of the Arya Samaj—the society of the Aryan race. Underlining the Brahmin perspective he stated that the Aryans were linguistically and racially pure and migrated from Tibet into India (Sarasvati 1935). He claimed that they established the purest and finest culture in India and that this should be revived. The upper castes were the Aryans from whose culture the untouchables/Dalits were excluded. The latter could be incorporated once they had gone through a purification ritual. There were some common ideas between the Arya Samaj and the Theosophical Society, also active particularly in South India, during the later 19th century.

In the early 20th century there was another shift in the identity of the Aryan. H.V. Savarkar and M.S. Golwalkar changed the identification from caste to include religion. The Hindus were now defined not only as the primary citizens of India, but also as the Aryans in the Indian population. All others were aliens. The term Aryan was now given a religious connotation. India (i.e., the territory of British India), had to be the land of one's ancestors, *pitribhumi*, and the land where one's religion originated, *punyabhumi*. Therefore, only the Hindus were eligible. This was to be a foundational argument to the more extreme Hindu nationalism of the 20th century, often labeled as Hindutva. The concept of the Aryan was now getting enmeshed in a variety of interpretations and definitions, very different from its earlier meaning.

The discovery of the Indus civilization/Harappa culture in the 1920s with a script, still unread, changed the picture. The Aryans were no longer the bedrock of Indian history. The beginnings now went back to the Indus civilization. What was its identity? Is it different from the agropastoral society of the Vedic texts? The nuclei were the cities and these were unknown to the Vedas. Because the language has remained unread, claims have been made both by supporters of Indo-Aryan and proto-Dra-

vidian viewpoints. One solution to the problem was to maintain that the Indus civilization was identical to the Vedic and represented its archaeological counterpart. This was first suggested by L.A. Wadell in 1925 when not much was known about the cities. The Hindutva ideologues in recent decades have been saying much the same but encountering opposition from many archaeologists and historians who find it unacceptable in the absence of evidence.

The latter had generally distanced themselves from variants of the theory. The initial argument had been that the Harappan cities declined owing to the invasion of the Aryans. This was questioned for lack of evidence and the decline traced to environmental factors. Extensive archaeological work in the Indus plain and adjoining areas revealed a large number of settlements of varied archaeological cultures, contemporary with the Harappan cities. Some of these continued into post-Harappan times. This has led to a reevaluation of the process by which the Indo-Aryan language spread in northern India. It seems more likely that there were small-scale migrations into the northwest and settlements within the vicinity of earlier settlements or even merging with these. The *Rigveda*, the earliest of the Vedic texts, is generally dated to between 1500 and 1000 BC. It shows linguistic elements and vocabulary from Dravidian and Munda languages, which may indicate bilingualism, which in turn would suggest a fair degree of mixing of the populations (Kuiper 1991; Witzel 1999). This would come about through intermarriage and the assimilation of each other's patterns of living.

Scholars have argued that even when cultures decline, there is always the possibility of some of their myths, rituals, and belief systems being continued through the oral tradition. This relates to the question of which of the two cultures was earlier. Archaeology generally provides reliable dates. Those for the Harappan cities point to a beginning in about 2650 BC for the cities in the northwest and a slightly later start for those in western India. The decline comes about a thousand years later in approximately 1750 BC. Textual evidence is less easy to date. The date of the *Rigveda* is tied to evidence from elsewhere, such as cognate words and concepts in the *Avesta* from Iran, and the names of deities in the Mitanni-Hittite treaty of the 14th century BC. The treaty has archaic forms of Indo-Aryan and could therefore probably be a little older than the *Rigveda*. This would in fact date the *Rigveda* to later than 1500 BC. In any case, it cannot be contemporary with the Indus civilization. The sophisticated urban culture of the Indus cities is not reflected in the agropastoralism of the *Rigveda*. The question of chronology assumes centrality if the Harappans are to be described as Aryans as some are trying to do. This is more an argument of contemporary political ideology rather than that of an analysis of the textual and archaeological evidence.

Much of this particular argument results from a concern to determine the indigenous and the foreign groups in the population. But we must remember that there were no cartographic boundaries in those days. The boundaries of British India, which are the ones used in these discussions, were the much later creation of British colonialism. In the precolonial period, the effective boundaries were based on common languages, practices, customs, and political control, the last of which was blurred at the edges. These were often assumed and not necessarily well defined. In these circumstances, groups can only be defined as conforming, to a degree, to a particular culture rather than being clearly demarcated as either indigenous or foreign.

I have tried to trace the mutations of the term arya and the notion of Aryan as an exclusive category. Its application has changed over the centuries and new turns of meanings have been introduced even in the way it has been used by colonial and nationalist scholars. Clearly the connotation of race does not apply. To argue that they were an exclusive and self-perpetuating group is not meaningful. Historically, it makes greater sense to recognize that from the second millennium onward there has been a mixing of the descendants of the Harappans living in various parts of the north and the west with other populations of the time and subsequently with those that have migrated into the subcontinent at various times, starting

with the Indo-Aryan speakers. The borderlands, in particular, had mixed populations. Some of these peoples then moved eastward into the Ganges valley with further mixing with the local populations there. The same procedure was followed in the movement southward into the peninsula. The degree of mixing cannot be ascertained, but linguistic traces of other languages in a given language can be used as a cautionary gauge.

There is now a turning to genetic data as a source for monitoring migrant and immigrant groups and arguing for a clear-cut descent of some of these. The intention within India is also to apply the results to ascertaining the identity of the Aryans. But as I have tried to show, Aryan is a social construct and therefore genetic information is unlikely to be useful unless the parameters defining the groups for analysis undergo some rethinking. Genetic data and analyses through the procedure of collecting and classifying samples may have to consider alternative criteria. Because the data involve social history, there is a very long span of time not only for new variation but also for assimilation from elsewhere.

Genetic profiles often assume that the claims of varna/caste categories as unchanging and exclusive are correct. But being socially created with no inherent natural rules, the identities of varna have undergone change. In the Vedic texts there is evidence of this. The most striking is the category called the dasi-putra Brahmanas. These are Brahmins identified as the sons of dasis. The word could mean either women of the dasa group or else a slave woman who would be of the lowest caste. In either case, the label contradicts the high status of the Brahmana claiming purity of descent. We are told that they were at first reviled by the regular Brahmins but when their superior ritual power was revealed they were eagerly assimilated into the orthodoxy. Their progeny would have been Brahmins but with a mixed ancestry. Among the better known in this category was the much-respected seer, Kakshivant (*Brihad-devata* 4.11-15; *Aitareya Brahmana* 8.23). A similar kind of recruitment to the Brahmana varna takes place in later centuries with the spread of Sanskritic courtly culture to outlying areas. Priests from local jatis who picked up a smattering of Sanskrit were recruited to perform the rituals. Their composition of Sanskrit inscriptions shows their scant knowledge. But over a few generations the family would become proficient and claim to be of the high status Brahmin varna.

Another varna thought to be concerned with purity of descent is that of the Kshatriyas. As the warrior aristocracy of epic times, and the multitude of rulers that were scattered across the subcontinent from the late first millennium AD, they were also viewed as a caste that preserved coherence over long periods. But, as it turns out, they are among the most open groups, perhaps because political power was in itself open. References to Kshatriyas occur in the later Vedic texts and in the *Puranas* (Wilson 1961). In both, there are long lists of succession and descent. Yet, even the most respected Kshatriya among them, Puru, the son of Yayati and ancestor to the protagonists of the *Mahabharata*, is said to have faulty speech, *mridhra-vac* (which would disqualify him as an arya), and descended from the demons, *asura rakshasas* (a serious disqualification of lineage).

In the late first millennium AD, adventurers of obscure origin or feudatories who rose to independent status, would, on setting up independent kingdoms, claim to be Kshatriyas. Support came from Brahmins who prepared genealogies for them, which occasionally were fabricated, and performed the necessary rituals for their legitimation as Kshatriyas, in return for a fee, which was often a grant of land. Such grants became the nucleus of those Brahmin families who a few generations down claimed to be independent rulers of the lands granted, and these families constituted a new category of *brahma-kshatra*, a mix of both (Fleet 1970). Mention is made in the *Puranas* of kings creating a caste of new Kshatriyas, sometimes from among those regarded as outside the pale of caste society. Dynasties of the period from ca. 500 BC to AD 400 are said to be of either the Shudra or Brahmin caste. Foreign rulers, such as the Indo-Greeks, are described as vratya or degenerate Kshatriya. The category of Kshatriya was open to various castes that might wish to claim it, as happened

from the latter part of the first millennium AD. Those claiming this status merely had to show their power in the open arena of politics.

The Shudra caste, as the lowest of the four, was also known to take on a variety of professions, from artisans to kingship. Those that ran the administration in many kingdoms were Kayasthas who are said to be of mixed castes, sometimes linked to one Shudra parent. Literacy was at a premium among these groups. Kayasthas in some instances are known to be authors of Sanskrit texts and as brilliant as their Brahmin counterparts.

It would seem therefore that to take varna as a consistent and controlled genetic group can be contrary to historical data. Nor can the caste codes as outlined in normative texts—the *Dharmashastras*—be taken literally. These were the idealized norms but obviously they were not widely observed, although the varna labels continued to be used. In fact, the jati structure of society is more reflective of social reality. But this network is perhaps too complex to be correctly followed for genetic analyses.

Categories of population groups are sometimes used as units of social analysis. But populations do not remain in one place for all time, and some among them frequently migrate. This is more characteristic of pastoralists in search of pastures and water supply than of sedentary agriculturists. But the latter are also dependent on fertile soil and water and therefore are known to have moved to many parts of the subcontinent in search of better conditions. The desiccation of the Indus plain led to migrations into the Ganges plain and from there in time to western and eastern India. In both of these areas there were prior populations of agriculturalists with whom the newcomers would have intermarried and have adopted some of their cultural traits. Encroachments into forested areas, increased from the late first millennium AD, would also have involved interaction with the forest dwellers. The frequency of Apsarases, celestial maidens, entering the genealogies, is suggestive of marriage with women outside the caste. Activities linked to trade, by their very nature require travel and habitats in areas beyond the usual, with some degree of intermarriage.

All this creates a mixing of populations. The border areas of the subcontinent have been home to a variety of peoples. The northwestern borders received large groups of migrants from Central Asia, Iran, and Afghanistan. The northeast became home to groups coming in from Tibet, China, and Burma. The ritual specialists and the more learned members of the Brahmin varna were particularly mobile from the mid-first millennium AD. So too were the low status jatis of architects and sculptors. Architectural styles in the royal temples of the Himalayan kingdoms reflect the hand of artisans from distant parts of the plains. This met the demand for specialists of various kinds in the new kingdoms. Brahmins from Gauda (eastern India), Kanauj (central Ganges plain), and Kashmir were employed in courts distant from their homes. Some among these established new gotras and pravaras, subdivisions of the caste, the names of which suggest recruitment from local jatis.

Another category suggested for identifying groups is language. If this is used, then precision can only be assumed for current language speakers. Languages also migrate and travel and are used by speakers other than the original. Those associated with ritual and with social and political status, such as Sanskrit, are often the most sought after. The spread of Sanskrit to South India, which was a Tamil-speaking area, indicates elite groups seeking status through adopting this language, quite apart from its use in ritual. This also raises the question of how a group is to be defined through language because language also reflects social hierarchy. For example, how do we define speakers of English in India? Studies of language can, however, provide some clues. The migration of a language or its intersection with other languages can be traced to an approximate extent through comparative studies of syntax, morphology, and phonetics—methods used in linguistics, but the methods have to be used with careful controls.

Social isolation and containment is not associated with castes, but with those that were excluded. It might be more useful to consider the DNA pattern among Scheduled Castes who have been regarded as untouchables throughout

history and whose mixing with other groups was therefore limited. These groups, now referred to as Dalits, were isolated because they were thought by Brahmins to be polluting. They had their own social code distinct from the rest. Other isolated groups were the Scheduled Tribes whose habitat in forests encouraged their isolation. The best case of this would be the Jaravas of the Andamans. Others such as Angami Nagas and Muria Gonds may have ceased to be isolated some time ago. Comparative studies of such groups both among themselves and in relation to caste groups might be useful. The lower end of the social spectrum is likely to be more meaningful in studies of DNA from a historical perspective.

For the historian, and specifically the historian of ancient India, there are problems with using the results of what is termed ancient DNA. Samples collected from archaeological sites can be suspect for a variety of reasons. They can be contaminated through lying for a few thousand years in the soil, as in the case of burials, or in porous containers such as urns or wooden coffins. There can be decomposition and the growth of bacteria that would affect the result. Also, DNA gets chemically modified over time. It would seem that ancient DNA might require new techniques in collecting samples and analyzing them. Techniques are, however, improving and more reliable forms of assessing the evidence may well emerge.

Of the genetic studies performed so far, the results have been contradictory and historians at least find it difficult to use them with any certitude. The debate now involves not only ascertaining whether some groups were indigenous and others migrants but also the degree to which there were mixed communities. The question of whether "the Aryans" were indigenous to India or which of the castes should be regarded as inheritors of the land, are questions that touch on current political ideologies. There is a tendency to pick up only those results that suit a particular theory. It could be argued that even if answers are found to such questions, these would not necessarily clarify the picture of what happened in the ancient period of Indian history. The questions generally posed to

Cite this article as *Cold Spring Harb Perspect Biol* doi: 10.1101/cshperspect.a008599

genetic data have largely been based on confirming conventional views of the flow of social history. The contradictions that might emerge could be used to search for a different kind of analysis.

Mine is not an attempt to resist genetic analysis of populations in India. I would, however, maintain that where identities based on exclusions or differentiations are used, it is important to recognize that these are not naturally given but socially constructed. The social constructions also have to be taken into account, and if need be, the units of analysis have to be readjusted. One must remember that there was a time when science was believed to have certitude and that "race science" drew on this. Science has certitude within the parameters of known knowledge, but when the parameters change the certitude also changes and has to be adjusted to the new parameters.

Given the evidence, the defining of social groups for DNA analysis should perhaps be reconsidered instead of being restricted to race, caste, and language. So too, the techniques of examining the data from the past could perhaps be refined further.

REFERENCES

Das AC. 1920. *Rigvedic India*. Cosmo Publications, Delhi, India (reprinted in 1987).

Deshpande GP, ed. 2002. *Selected works of Jyotirao Phule*, pp. 23–100. Leftward, Delhi, India.

Dolgin E. 2009. The history of two distinct lineages led to most modern-day Indians. *Nature* doi: 10.1038/news.2009.935.

Fleet JF, ed. 1970. Khoh copper-plate inscription of Maharaja Hastin. In *Corpus inscriptionum indicarum (1888)*, Vol. III. Indological Book House, Varanasi, India.

Kuiper PBJ. 1991. Aryans in the *Rigveda*. *Leiden studies in Indo-European, I*. Rodopi, Amsterdam, The Netherlands.

Manu. 1991. *Dharmashastra*, pp. 10, 45, 57, 66–73. Penguin, Delhi, India (translated by W. Doniger and B. Smith).

Poliakov L. 1974. *The Aryan myth: A history of racist and nationalist ideas in Europe*. Basic Books, New York.

Ramaswamy S. 1997. *Passion of tongues*. University of California Press, Berkeley, CA.

Risley HH, Crooke E. 1915. *The people of India*. W. Thacker, London.

Sarasvati D. 1935. *Satyartha prakash (Light of truth)*. Sarvadeshin Arya Pratinidhi Sabha, New Delhi, India.

Schwab R. 1984. *The oriental renaissance: Europe's discovery of India and the East (1680–1880)*. Columbia University Press, New York.

Staal F. 2008. *Discovering the Vedas*. Penguin, Delhi, India.

Thapar R. 2008. The *Rigveda*: Encapsulating social change. In *The Aryan: Recasting constructs*. Three Essays Collective, Delhi, India.

Trautmann TR. 1997. *Aryan and British India*. Oxford University Press, New Delhi, India.

Trautmann TR. 2006. *Languages and nations: The Dravidian roof in colonial Madras*. University of California Press, Berkeley, CA.

Wadell LA. 1925. *The Indo-Sumerian seals deciphered*. Octavo, London.

Wilson HH, ed. 1961. *The Vishnu Purana: A system of Hindu mythology and tradition, Book IV*. Punthi Pustak, Calcutta, India.

Witzel M. 1999. Substrate languages in old Indo-Aryan (Rigvedic, middle and later Vedic). *Electron J Vedic Stud* **5:** 1–97.

Race in Biological and Biomedical Research

Richard S. Cooper

Department of Public Health Sciences, Loyola University Medical School, Maywood, Illinois 60153

Correspondence: rcooper@lumc.edu

The concept of race has had a significant influence on research in human biology since the early 19th century. But race was given its meaning and social impact in the political sphere and subsequently intervened in science as a foreign concept, not grounded in the dominant empiricism of modern biology. The uses of race in science were therefore often disruptive and controversial; at times, science had to be retrofitted to accommodate race, and science in turn was often used to explain and justify race. This relationship was unstable in large part because race was about a phenomenon that could not be observed directly, being based on claims about the structure and function of genomic DNA. Over time, this relationship has been characterized by distinct phases, evolving from the inference of genetic effects based on the observed phenotype to the measurement of base-pair variation in DNA. Despite this fundamental advance in methodology, liabilities imposed by the dual political-empirical origins of race persist. On the one hand, an optimistic prediction can be made that just as geology made it possible to overturn the myth of the recent creation of the earth and evolution told us where the living world came from, molecular genetics will end the use of race in biology. At the same time, because race is fundamentally a political and not a scientific idea, it is possible that only a political intervention will relieve us of the burden of race.

A Klee painting named Angelus Novus shows an angel looking as though he is about to move away from something he is fixedly contemplating. His eyes are staring, his mouth is open, his wings are spread. His face is turned toward the past. Where we perceive a chain of events, he sees one single catastrophe, which keeps piling wreckage upon wreckage and hurls it in front of his feet. The angel would like to stay, awaken the dead, and make whole what has been smashed. But a storm is blowing from Paradise; it has got caught in his wings with such violence that he cannot close them. The storm irresistibly propels him into the future to which his back is turned, while the pile of debris in front of him grows skyward. This storm is what we call progress.

—Walter Benjamin
Theses on the Philosophy of History

RACE AS THE UNWELCOME GUEST IN THE DISCIPLINES OF SCIENCE

We rarely appreciate the presence of history in our day-to-day experience. The quotidian is a mixture of the repetitive and the predictable, carried forward by habit and punctuated by random events that we regard as either good or bad fortune. But in a more reflective mood, we have to acknowledge the relentless force of history that holds us in its grasp and accept that it creates the possibilities we use to negotiate with the future. The imposition of racial categories on human populations has been one of the most enduring historical forces that sets limits on opportunity and thereby shapes our life trajectory. As a projection of the underlying power relationships onto individuals, racial categories are used to structure social inequality. These power relationships are manifested both in the belief system that rank orders intrinsic human qualities according to group membership and the social institutions that enforce this hierarchy by restricting access to wealth, education, and other social goods. This daily reality is central to the history of all modern societies.

The racial structuring of society also has pervasive influence on biological research and the patterns of health and disease. Enormous effort has been expended to describe human demographic history through reference to an ever-changing array of constructs and categories, all of which include a hierarchical arrangement—either explicit or implicit. In the United States, most prominently, public health has embraced racial/ethnic categories as fundamental structural elements. Clinical medicine has similarly evoked racial categories to explain causation and outcomes across the entire spectrum of diseases. At the same time, race has met some of the strongest challenges to its legitimacy in biology and biomedicine. All of biology is grounded in the theory of descent from a common ancestor. The belief in racial categories was one of the most powerful liabilities of premodern biology and lent credence to the established view of divine creation. Indeed, it has recently been argued that the challenge to race brought by the abolitionist movement was a key factor behind Darwin's transformative insight that the biological world—on the evolutionary time scale—is a single indivisible whole (Desmond and Moore 2009). Biomedicine still grapples with the implications of that insight for our species, yet substantial progress—uneven, tentative, and ultimately disappointing—has been made. In the current era, genomic science has opened new vistas onto previously unobserved dimensions of biology, and that proportion of the concept of race that has been attributed to genetics can finally be subjected to empirical scrutiny. Integrating this new knowledge into practice and focusing the technology on socially productive work, as always, remains our most difficult challenge.

The narrative of race therefore wanders the border territory between what we call science and what we recognize as history and politics. In the pregenomic era, there was no requirement—indeed, no opportunity—to validate the authority of race with molecular evidence; causal inferences were made on the basis of phenotype, in its broadest possible sense, from disease to accumulated material wealth to social graces. The primary purpose of the race concept was to serve as a shortcut, an organizing tool that allowed postenlightenment Europe to explain—and thereby justify—how imperialism had reshaped the world. Consequently, for both the social and biological sciences, race felt like the rude cousin whose claim on our affection was based on obligation, not choice. In every historical period, an incremental struggle has been waged to overcome the disruption that this unwelcome intruder has caused within empirical scientific disciplines.

In its origins, race was a "label of convenience" that biologists used interchangeably with the construct of "varieties" as they tried to create taxonomic categories below the level of the species (Cooper 1984). Writers from across the intellectual spectrum of literature and politics also felt free to make use of the idea. Thus, Baudelaire spoke of the "race of Abel" and the "race of Cain" when describing the polarization of 19th-century French society, and Marx characterized the English working class as a "race of peculiar commodity owners" (Baudelaire 1857; Marx 1957) ("Hence the sum of the means of subsistence necessary for the production of labour-power must include the means necessary for the labourer's substitutes, i.e., his children, in order that this race of peculiar commodity-owners may perpetuate its appearance in the market." [p. 172]). Malleability continues to be an essential quality of race, although it is now primarily used as a label for the temporary and often random aggregation of population subgroups, usually tied in some rough way to the perceived continent of origin (Kaufman and Cooper 1996). In its contemporary sense, biological race has now come to signify the inherited qualities of a population group hidden inside the DNA molecule.

 Cite this article as *Cold Spring Harb Perspect Med* doi: 10.1101/cshperspect.a008573

RACE AS A DEVICE TO UNDERSTAND THE BIOLOGICAL WORLD

The uses of race in biomedical research in the United States can be divided roughly into four historical periods. From the mid-1800s the dogma of racial inferiority meant that racial-genetic explanations were invoked as biological justification for discriminatory and genocidal policies (Montagu 1942; Cooper 1993a). The sea change that resulted from antifascist struggles of the 1930s and 1940s, together with emergent voices of African-American self-assertion, issued in a second phase and constrained the use of blatant biological determinist arguments. In the last half of the 20th century and extending into the present, the utility of race has been viewed from two distinct perspectives: as a descriptive category—necessary to document health inequalities—and as a causal explanation of ill health—through unspecified genetic influences (Cooper 1993b). Finally, in the contemporary era, the standard has shifted to require molecular evidence for causal effects that are to be ascribed to biological race; if this standard is not always met in practice, it can now at least be justified in theory.

The terms of the debate over "race and health" have therefore been sequentially recast. The first major challenge to claims of genetic causation was directed toward the "inference from phenotype" approach. Because most common diseases result from chronic exposures to noxious environmental stimuli, and because the pattern of exposure to the entire gamut of environmental exposures is highly structured by social class, geography, and other historical factors, disentangling an "intrinsic" property of race from the summed effect of poorly measured or unknown external risk factors is generally not feasible. In technical terms, because "race"—as a social category—is designated with relative precision, and lifetime exposure to the wide range of correlated risk exposures is characterized only approximately for individuals, causal analyses are undermined by residual confounding (for a more detailed discussion, see Kaufman et al. 1997; Cooper 2001). The "race" variable, and its imputed biological meaning, retains statistical significance that should most likely be distributed over other factors in the analytic model, or factors that remain unmeasured.

Typically, in studies comparing U.S. blacks and whites, income and education are the primary variables used to adjust for the influence of the social environment. On reflection, however, it is apparent that a whole universe of other factors, e.g., quality of education, neighborhood of residence, inherited wealth, interaction with the criminal justice system, etc., have a profound impact on life's chances and physical and mental well-being. Although the logic of the argument for residual confounding has been generally accepted by epidemiologists, in practice it is more often ignored—perhaps in part because it identifies a problem without offering a solution. In fact, we will almost never have the capacity to summarize mathematically the impact of racial discrimination. Nietzsche recognized the philosophical implications of this problem, which is deeply rooted in many forms of belief: "The falseness of an opinion is not for us any objection to it. . . The question is, how far an opinion is life-furthering, life-preserving, . . .and we are fundamentally inclined to maintain that the falsest opinions are indispensable to us, that without a recognition of logical fictions, without a comparison of reality with the purely imagined world of the absolute and immutable, without a constant counterfeiting of the world by means of numbers, man could not live" (Beyond Good and Evil, Chap. 1, epigram 3). Although usually interpreted as a criticism of Platonism and traditional religion, this characterization of useful fiction is equally apt when applied to belief in racial superiority, and it is precisely the "counterfeiting. . .by means of numbers" that concerns us here.

The ability to probe the genome has now changed the course of much of human biology, including such disciplines as evolution, epidemiology, demography, and clinical medicine. Genomics is also fundamentally altering how race is conceptualized and used in biomedical research. In the initial phases, the dominant constructs of the pregenomics era were simply carried forward. Large numbers of candidate gene studies, for example, purporting to show a higher frequency of "risk alleles" in populations with increased disease rates—usually African Americans—were published from multiple disciplines. These studies were often interpreted as offering an explanation of public health inequalities. On technical grounds alone, this generation of candidate gene studies was ill conceived as a way of elucidating heritable influences and in fact yielded mountains of false-positive results; it was therefore easy to exploit this approach to satisfy the inherent biases of race. Fortunately, genome-wide association studies have proven to be more reliable, and for the first time we have credible insights into variation in genetic factors across a range of populations.

A vast amount of new information is being acquired with the tools now available, and although we are still in the early stages, for the first time the genetic architecture of many health-related traits is being described. As a consequence, we also now have the opportunity to accumulate data on the overarching question of the contribution of genes to racial patterns of disease. In parallel with this progress on disease-associated genetic variation, much deeper insight has been obtained into population genetics, quantifying relatedness and difference among racial/ethnic groupings.

All of the major causes of disability and death in the United States occur more frequently in blacks than in whites, with the exception of lung disease and suicide. Although all of these conditions have been analyzed from the basis of racial predisposition, none have been more central to the discourse about race and health in North America than hypertension. As a disorder that results from "dysregulation" of a basic physiologic system, hypertension has an ill-defined status, putting it somewhere between a "trait" and a "disease." It cannot be biopsied, X rayed, or localized to an organ or biochemical pathway. Conceptually, it reflects a free-floating quality of the whole organism, and its origin is therefore subject to a wide range of speculation, in many ways mirroring the essentialist attribute of race. This quality may underlie some of the unyielding temptation to ascribe higher risk of hypertension among blacks to a

genetic cause (Cooper and Rotimi 1994; Cooper et al. 1999).

HIGH BLOOD PRESSURE AND AFRICANS IN THE NEW WORLD

By the 1930s, both clinical and population studies had shown that black Americans had higher average blood pressure than did persons of European descent (Cooper et al. 2000). The sequelae of hypertension—primarily stroke, heart disease, and renal failure—still account for most of the overall health disadvantage experienced by black adults in the United States. Among blacks older than age 55 in rural Georgia, mean systolic blood pressures approaching 180 mmHg were recorded in the 1950s. Because the researchers were unable to identify causal factors in the environment, the prevailing view was that "the frequency of hypertension in a given population. . .(was determined by). . .the frequency in that population of a hypertensive gene" (White 1967, p. 183). Paul Dudley White, the preeminent American cardiologist of the first half of the 20th century, was particularly taken by the "high frequency with which hypertension has been found to exist among our American Negroes" (White 1967, p. 150). Once we are able to forgive the quaint manner of speech that White used to frame the question, it is undeniable that he formulated an important insight into this problem: "I used to ask my medical friends. . . about the situation in West Africa as to the prevalence of. . .hypertension in the jungle. . .." (White 1967, p. 150). But the evidence that White could rely on at that time was fragmentary, based on hospital cases from what was then Leopoldville in the Congo, and he concurred with the prevailing notion of "racial predisposition" (White 1967, p. 151). This perspective remains the dominant view until today.

As a medical student in the late 1960s, I too was deeply moved by the severity of hypertension among black patients I saw from the delta farmlands of eastern Arkansas. Devastating strokes in patients in their 40s and 50s were common. In the context of the 1960s, especially in Arkansas, there was good reason to question all

 Cite this article as *Cold Spring Harb Perspect Med* doi: 10.1101/cshperspect.a008573

assumptions about race. The invocation of a mysterious essentialist quantity inherent in an individual with darker skin color—in the face of profound differences in life experience—seemed like nothing more than another way of formulating the "big lie" about the history of the south. At that time, of course, there was no way to test the validity of any of the conventional claims being made. In fact, through the 1990s, I still maintained that attempts to define the mechanism underlying black–white differences in blood pressure belonged in the realm of ne-science—the study of the unknown and the un-knowable—rather than science (Cooper and Kaufman 1998).

As a result of interactions with colleagues from West Africa and the Caribbean, I was presented with the opportunity to test at least the "inference by phenotype" version of the black–white blood pressure story. Using neighborhood samples, my colleagues and I recruited 11,000 adults between the ages of 25 and 64 in urban and rural West Africa (Nigeria and Cameroon), the West Indies (Barbados, St. Lucia, and Jamaica) and the United States (metropol-

itan Chicago). Blood pressures in rural Africa were low and rose very little with age; in the West Indies, average blood pressures were the same as those found among whites in North America; the metropolitan Chicago sample (drawn from the community of Maywood), on the other hand, reproduced the U.S. data as a whole, with a 50% excess of hypertension, compared with whites (Fig. 1) (Cooper et al. 1997). As one might expect, the lifestyle factors known to increase blood pressure—namely, obesity, high intake of sodium, and low intake of potassium, fruits, and vegetables—paralleled the gradient in blood pressure. Subsequently, we compared these data to national surveys in the United States, Canada, and several European countries (Fig. 2) (Wolf-Maier et al. 2003). The "white" populations of Europe had substantially higher blood pressures than the non-U.S. Afro-origin populations, and in northern Europe levels were similar to or higher than that of the U.S. blacks study population (Cooper et al. 2005). Other studies have subsequently shown that immigrants to the United States from the Caribbean have lower blood pressures and lower stroke

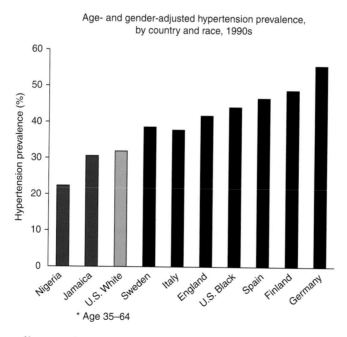

Figure 1. Prevalence of hypertension among six populations of West African origin. (From Cooper et al. 1997; reprinted, with permission, from the author.)

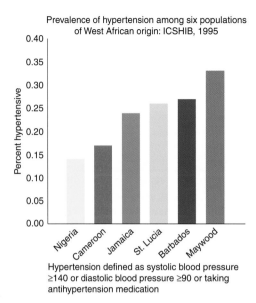

Prevalence of hypertension among six populations of West African origin: ICSHIB, 1995

Hypertension defined as systolic blood pressure ≥140 or diastolic blood pressure ≥90 or taking antihypertension medication

Figure 2. Age- and gender-adjusted hypertension prevalence by country and race in the 1990s. (From Cooper et al. 2005; reprinted, with permission, from the author.)

rates than U.S.-born blacks. Clearly, therefore, the construct of "black exceptionalism" in terms of the risk of hypertension must be reconsidered as at most a "permissive susceptibility." The true counterfactual state—persons of recent African descent who live the U.S. lifestyle without facing either racial or class disadvantage—does not exist, and although deprived of any affirmative evidence, the answer to the question of genetic predisposition remains unknown.

GENOMICS AND STUDIES OF RACIAL FACTORS IN DISEASE

Conclusions about racial predisposition based on the observation of difference in phenotype have therefore become untenable—although they continue to be routinely reported in some fields—and the complex task of sorting through the molecular evidence has begun (Cooper et al. 2003). As described elsewhere in this collection, one of the first constructs that required revision was the notion of "unit race"—that the various lists of population groups were cataloging similar segments of the world's population. The

evolutionary history of *Homo sapiens* was immediately apparent in any genetic studies that included African-origin groups. Our family tree is rooted in Africa, and the vast majority of accumulated genetic variation is present in contemporary African populations. Thus, Africans are no longer the "outlier" or the "other," but in fact they are the source population and the reference point for all other continental groups.

As suggested above, the application of this new technology to the study of common disease followed a confused path through family studies, comparison of case and noncases at specific loci, and finally genome-wide association studies simultaneously comparing markers on all chromosomes. Statistical genetics had long since provided the basis for a strong assumption that quantitative traits were multigenic, that is, many mutations in many genes and their regulatory components were responsible for interindividual variation in phenotypes. Surprisingly, geneticists and epidemiologists alike opened this chapter of research with the opposite assumption that mutations of relatively large effects could be found. When integrated into a search for genetic factors underlying racial predisposition, the choice of assumption is obviously crucial. In modern societies, chronic diseases are the primary threat to health. Two-thirds of all deaths result from cardiovascular diseases and cancer alone. The pathophysiology of these conditions as a rule reflects dysregulation of metabolic pathways (e.g., lipid metabolism, blood pressure control, and DNA repair mechanisms). Two important attributes of these physiologic systems make them unlikely candidates for the "large effects–few genes" scenario. First, because these pathways are deeply embedded in normal physiology, they have been finely tuned by evolutionary selection and are controlled by many genes. Second, none of the specific phenotypes manifested as disease states would have been subject to positive selection because they occur with aging and have little impact on reproductive fitness. At the present time, strong evidence of selection in the human genome is restricted almost exclusively to loci that protect against epidemic infectious processes of relatively recent origin, with malaria still the best-

 Cite this article as *Cold Spring Harb Perspect Med* doi: 10.1101/cshperspect.a008573

described example, or the ability to extract nutrient value from particular foods such as milk. Selection is almost always required to drive large interpopulation differences in functional genetic variants.

The search for racial differences in genetic predisposition must therefore be framed by this evolutionary context. If genetic susceptibility reflects multiple mutations in many genes that have no impact on fitness, it is highly improbable on statistical grounds that large numbers of these mutations would aggregate in one geographic population. This is even less probable for Africans, who share the vast majority of all variation in our species. Thus, within Africa, the entire spectrum of many phenotypic traits is observed (e.g., among the tallest populations and the shortest populations or groups with intestinal enzymes that allow metabolism of milk products and those without).

Although the genetic architecture underlying blood pressure regulation is still very poorly understood, the physiology makes it clear that many genes must be involved. The instantaneous response of blood pressure to the metabolic demands of every organ is an essential requirement for fitness among all animals with a cardiovascular system. In addition to being highly redundant and finely tuned, blood pressure regulation is a coordinated effort of multiple pathways—the nervous system, the heart, the kidneys, and the hypothalamic-pituitary-adrenal axis, among others. We now have empirical evidence that height is likely to be influenced by 4–500 genetic loci (Lanktree et al. 2011); it would be surprising if any fewer accounted for the distribution of blood pressure in the population. Large studies on blood pressure have failed to identify loci that have major effects, and although this may be in part owing to the difficulty in characterizing "average" blood pressure of an individual, it shows that genes with large effects are not common in human populations. For example, a study of 200,000 participants documented only a handful of genetic loci that were significantly associated with hypertension or blood pressure (The International Collaboration for Blood Pressure Genome-Wide Association Studies 2011). A combination of the 29

markers that had the strongest association with blood pressure raised the risk of hypertension by only 23%. Given that the lifetime risk of hypertension in the United States approaches 85%, the genetic information offers little predictive value so far. If, as expected, the distribution of genes with susceptibility alleles, and the alleles themselves, also vary to some degree among populations, and interactions with the environment are important, it will be extremely difficult, perhaps impossible, in the foreseeable future to calculate some net "population risk score" for multigenic traits such as hypertension.

Based on the evidence now available from studies of traits including height, we can model the underlying distribution of genetic variants that are common in populations (common meaning occurring in >5% of affected individuals). As seen in the accompanying figure, occasional genetic variants have been found that influence the complex traits by as much as 35%; however, these are very few in number (see smoothed estimate of distribution of regression coefficients associated with minor alleles for susceptibility loci, shown with red and without [blue] power adjustment in Park et al. [2011]). The underlying distribution is most likely a truncated normal distribution (i.e., a very large number of mutations with small effects, most of which have still escaped detection). Assuming that this pattern is typical for most diseases and that these mutations occur at random, this evidence provides the best empirical support yet available for the argument that large shifts in risk by continental or other population grouping are unlikely.

At the same time, of course, exceptions have been observed. As noted previously, large differences among populations are described for mutations that are protective against epidemic infections. Two additional examples have now been found for chronic diseases that are more common in U.S. blacks than in U.S. whites. Prostate cancer occurs about 2.5 times as often in black men, and mutations that are associated with a substantial increase in risk are now known to be very common in populations of African— at least West African—descent (Chang et al.

2011). As this literature matures, however, the role of specific variants appears to be becoming more and more complex. In addition, a relatively uncommon form of chronic renal failure—focal segmental glomerulosclerosis—is strongly associated with a mutation in a gene related to lipid metabolism (*ApoL1*), and one copy of this version of the gene likewise occurs in almost 90% of West Africans (Rosset et al. 2011). Some evidence suggests that positive selection may be operating for the risk alleles for both prostate cancer and kidney disease—again, potentially through their role in immunity—although not all of the pieces of the puzzle fit together. As noted above, it remains exceedingly difficult to define the "causal" mutations that influence multigenic traits, and recent evidence suggests that *ApoL1* may not even be the target gene after all (S. Winkler, pers. comm.). However, if the original explanations are substantiated, these conditions would represent examples of race-specific susceptibility to common disease. Of course, a much more common condition based on race-specific genetic factors is present among Europeans who now live in sunny climates or at low latitudes, namely, their light skin. Needless to say, a host of other traits and diseases of smaller public health burden—from cystic fibrosis to sarcoidosis—also shows large variation among geographic populations. Sorting through the balance among geographic populations will become possible as the causal mutations are identified. A rough estimate of the contribution of known genetic variants to the overall health disparity between U.S. blacks and whites—summarized by life expectancy—yields a figure with an upper bound of 3% (Kaufman et al. 2013).

EMERGING FROM THE BELIEF IN RACE TO THE OBSERVATION OF DIFFERENCES

The contemporary experiment in large-scale genotyping has had contradictory effects on the meaning attributed to race in biomedicine. One major effect, certainly in the everyday perception of clinicians, has been a reinscribing of race as an important determinant of health and disease. Paradoxically, however, the massive outpouring of new knowledge about interindividual genetic variation has disposed of the proposition that race categories can serve as a broad proxy for genetic effects—both known and unknown. Furthermore, in many large cosmopolitan urban centers, the historical boundaries of race have blurred substantially. In New York City, for example, 36% of the residents were foreign born in 2005, speaking 170–200 languages. For the two largest official U.S. "minority" populations, blacks and Hispanics, a continuous spectrum of ancestral background is observed (Fig. 3) (Tayo et al. 2011); using ancestry information genetic markers, the figure depicts the proportion of ancestry from Europe and Africa

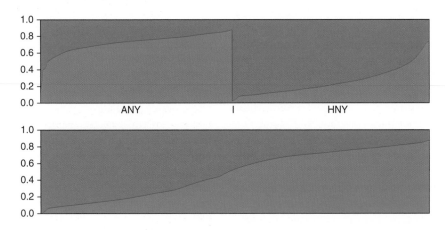

Figure 3. Proportion of African and European genetic ancestry in African Americans and Hispanics in New York City, 2008. Red, African ancestry; blue, European ancestry; ANY, African Americans in NY; HNY, Hispanics in NY.

 Cite this article as *Cold Spring Harb Perspect Med* doi: 10.1101/cshperspect.a008573

for each of the individuals studied. The standard racial/ethnic categories thus become less and less relevant in the genomic era and ultimately lose any value as a proxy for genotype.

The tension between the social prejudice regarding race and reasonable scientific evidence has thus been transformed. Given the opportunity to gradually sketch out the function of the genome and document its consequences for disease states, we can now begin to identify population-specific factors and assess their relative importance. Returning to the theme of historical context, it is worth noting that this tension has been a constant feature of modern biology since its origins. When he arrived in Paris in the 1860s to pursue the study of biology, Anténor Firmin, a Haitian intellectual and subsequent prime minister, was confronted with claims by the authorities in the field that "it is certain that the different races today are absolutely separate" (Firmin 1885). These claims were supported by evidence that he also had good reason to doubt, such as "the union of a Negro and a white woman are very often sterile." After a thorough study of the current state of knowledge, he published a book entitled *The Equality of the Human Races* using existing evidence to advance essentially the same argument outlined in this essay (Firmin 1885). As Firmin wrote, "Yes, human beings can and do differ by their physical traits or the color of their skin. Yet, they are equal in intelligence and thought. Only a long process of perversion of the spirit and very powerful influences on the minds of White people could have made them overlook a truth that is so obvious and natural that it requires no scientific proof." His hopeful prediction for the utility of biological science to improve the human condition is equally modern: "We still have a chance of eradicating this prejudice. . . . In the process, we will succeed in lowering the pretensions of an incomplete and ill-conceived science which continues unconsciously to validate the most hurtful errors through assertions that are as suspect as they are perverse." We can hardly lay claim to a better conceptual understanding than Firmin's, but with the advantage of molecular technology, we can now make progress toward a science that has matured by being rooted

in empirical evidence. Variation among geographic populations is real, and study of its origins can yield important biological insights. But there are no categories of race that segment human populations, and there are no mysterious qualities "in the blood" that justify the belief in racial superiority.

REFERENCES

Baudelaire C. 1857. *Abel et Cain: Les fleurs du mal* (ed. Pichois JZC, 1975). Gallimard, Paris.

Chang BL, Spangler E, Gallagher S, Haiman CA, Henderson B, Isaacs W, Benford ML, Kidd LR, Cooney K, Strom S, et al. 2011. Validation of genome-wide prostate cancer associations in men of African descent. *Cancer Epidemiol Biomarkers Prev* **20:** 23–32.

Cooper RS. 1984. A note on the biological concept of race and its application in epidemiological research. *Amer Heart J* **108:** 715–723.

Cooper R. 1993a. Use of race in public health surveillance: Perspective of a health scientist. *MMWR* **42:** 11–12.

Cooper RS. 1993b. Health and the social status of blacks in the United States. *Ann Epidemiol* **3:** 137–S144.

Cooper RS. 2001. Social inequalities, ethnicity and cardiovascular disease. *Int J Epidemiol* **30:** S48–S52.

Cooper RS, Kaufman JS. 1998. Race and hypertension: Science or nescience? *Hypertension* **32:** 813–816.

Cooper RS, Rotimi C. 1994. Hypertension in populations of West African origin: Is there a genetic predisposition? *J Hypertens* **12:** 215–227.

Cooper R, Rotimi C, Ataman S, McGee D, Osotimehin B, Kadiri S, Muna W, Kingue S, Fraser H, Forrester T, et al. 1997. Hypertension prevalence in seven populations of African origin. *Am J Public Health* **87:** 160–168.

Cooper RS, Rotimi CN, Ward R. 1999. The puzzle of hypertension in African Americans. *Sci American* **280:** 56–63.

Cooper R, Cutler J, Desvigne-Nickens P, Fortmann S, Friedman L, Havlik R, Hogelin G, Manolio T, Marler J, McGovern P, et al. 2000. Trends and disparities in coronary heart disease, stroke and other cardiovascular diseases in the United States: Findings of the National Conference on CVD Prevention. *Circulation* **102:** 3137–3147.

Cooper RS, Kaufman J, Ward R. 2003. Race and genomics. *N Engl J Med* **348:** 1166–1170.

Cooper RS, Wolf-Maier K, Adeyemo A, Banegas JR, Forrester T, Giampaoli S, Joffres M, Kastarinen M, Primatesta P, Stegmayr B, et al. 2005. An international comparative study of blood pressure in populations of European vs. African descent. *BMC Medicine* **3:** 2.

Desmond A, Moore J. 2009. *Darwin's sacred cause: Race, slavery and the quest for human origins.* Allen Lane, London.

Firmin A. 1885. *The equality of the human races* (reprinted by the University of Illinois Press, Chicago, 2002).

The International Collaboration for Blood Pressure Genome-Wide Association Studies. 2011. Common polymorphisms impacting blood pressure and cardiovascular disease in diverse populations highlight novel biological pathways. *Hum Mol Genet* **20:** 2273–2284.

Kaufman J, Cooper RS. 1996. Descriptive studies of racial differences in disease: In search of the hypothesis. *Publ Health Reports* **110:** 662–666.

Kaufman JS, Cooper RS, McGee D. 1997. Socioeconomic status and health in blacks and whites: The problem of residual confounding and the resiliency of race. *Epidemiology* **6:** 621–628.

Kaufman JS, Rushani D, Cooper RS. 2014. Nature versus nurture in the explanations for racial/ethnic health disparities: Parsing disparities in the era of genome-wide association studies. In *Reconsidering race: Global and comparative studies in race and genomics.* Oxford University Press, Oxford (to be published).

Lanktree MB, Guo Y, Murtaza M, Glessner JT, Bailey SD, Onland-Moret NC, Lettre G, Ongen H, Rajagopalan R, Johnson T, et al. 2011. Meta-analysis of dense gene-centric association studies reveals common and uncommon variants associated with height. *Am J Hum Genet* **88:** 6–18.

Marx K. 1957. *Capital,* Vol. 1, p. 172. Progress, Moscow.

Montagu A. 1942. *Man's most dangerous myth: The fallacy of race.* AltaMira, New York, 1997.

Park JH, Gail MH, Weinberg CR, Carroll RJ, Chung CC, Wang Z, Chanock SJ, Fraumeni JF Jr, Chatterjee N. 2011. Distribution of allele frequencies and effect sizes and their interrelationships for common genetic susceptibility variants. *Proc Natl Acad Sci* **108:** 18026–18031.

Rosset S, Tzur S, Behar DM, Wasser WG, Skorecki K. 2011. The population genetics of chronic kidney disease: Insights from the *MYH9-APOL1* locus. *Nat Rev Nephrol* **7:** 313–326.

Tayo BO, Thiel M, Tong L, Qin H, Khtrov G, Zhang W, Zhu X, Pereira A, Cooper RS, Bottinger EP. 2011. Genetic background of patients from a university medical center in Manhattan: Implications for personalized medicine. *PLoS ONE* **6:** e19166.

White PD. 1967. Hypertension and atherosclerosis in the Congo and in the Gabon. In *The epidemiology of hypertension* (ed. Stamler J, et al.), pp. 150–154. Grune & Stratton, New York.

Wolf-Maier K, Cooper RS, Banegas JR, Biampaoli S, Hense H, Joffres M, Kastarinen M, Poulter N, Primatesta P, Rodriguez-Artalejo F, et al. 2003. Hypertension and blood pressure level in six European countries, Canada, and the United States. *JAMA* **289:** 2363–2369.

Personalized Medicine and Human Genetic Diversity

Yi-Fan Lu[1], David B. Goldstein[1], Misha Angrist[2], and Gianpiero Cavalleri[3]

[1]Center for Human Genome Variation, Duke University, Durham, North Carolina 27708
[2]Institute for Genome Sciences and Policy, Duke University, Durham, North Carolina 27708
[3]Molecular and Cellular Therapeutics, Royal College of Surgeons, Dublin 4, Ireland

Correspondence: d.goldstein@duke.edu

Human genetic diversity has long been studied both to understand how genetic variation influences risk of disease and infer aspects of human evolutionary history. In this article, we review historical and contemporary views of human genetic diversity, the rare and common mutations implicated in human disease susceptibility, and the relevance of genetic diversity to personalized medicine. First, we describe the development of thought about diversity through the 20th century and through more modern studies including genome-wide association studies (GWAS) and next-generation sequencing. We introduce several examples, such as sickle cell anemia and Tay–Sachs disease that are caused by rare mutations and are more frequent in certain geographical populations, and common treatment responses that are caused by common variants, such as hepatitis C infection. We conclude with comments about the continued relevance of human genetic diversity in medical genetics and personalized medicine more generally.

We all differ at the level of our DNA sequence, and geneticists obsess over trying to understand the significance of this genetic diversity. This is an important goal, as by understanding human genetic diversity we can learn about the evolutionary history of our species, where we have come from, and perhaps where we are headed. More practically, understanding human genetic diversity is essential to understanding the biology of our diseases of various kinds, from the genetically more simple to more complex, and how we respond to treatment at both the population and individual levels (Torkamani et al. 2012). Indeed, improving our knowledge of human disease biology is the primary driver behind the largest and most systematic studies of human genetic diversity today. These studies, and the population- and disease-specific investigations made possible by them, are essential for reducing health disparities and improving health outcomes for the species as a whole. Unfortunately, largely because of which DNA samples are most easily accessible, most genomics research programs have concentrated their discovery efforts in populations of European ancestry (Need and Goldstein 2009;

Bustamante et al. 2011). As we discuss in this essay, this approach is myopic and carries with it untoward consequences for both the scientific and public health enterprises.

The successful completion of the Human Genome Project in 2003 was the first in a series of large multinational public efforts that began to move the field of medical genetics away from purely descriptive documentation of patients' physical features coupled with laborious one-by-one examination of a small subset of their genes for potentially pathogenic changes. For example, the International HapMap Project's collection of millions of genotypes from four global populations was indispensable to the pursuit of hereditary changes in genes that contribute to disease by providing the platform for so called "genome-wide association studies" (GWAS). GWAS gave us the ability to efficiently and comprehensively assay genetic variants that are common in a population and identify those that appear more commonly in patients with a given disease than they do in controls without the disease. Such variants can sometimes provide clues to the genetic basis of human disease (Manolio et al. 2009).

In parallel, researchers have capitalized on our improved understanding of population history to identify disease-causing genes. Population-specific studies of disease, from myocardial infarction in Icelanders to prostate cancer in African-Americans, have cleverly exploited the enrichment of specific disease-susceptibility alleles in more genetically homogeneous populations (Torkamani et al. 2012).

Along the way, advances in our understanding of patterns of human genetic variation have also informed our view of the history of modern human populations. Our interpretation of the scientific data, however, has been influenced by the constantly evolving sociopolitical milieu. During the early part of the 20th century, two schools of thought emerged on how natural selection influenced the frequency and distribution of genetic variation. The "classical" school believed that most genetic variation was rare and that variants present in the population are almost always deleterious. In the very occasional cases in which new mutations are advantageous, they quickly became "fixed" (cases in which the new advantageous allele replaced the ancestral one). The "balanced" school, on the other hand, believed that genetic variation was quite common and often actively maintained by selection favoring multiple forms of a gene in the population. This might be because of so-called overdominance, in which selection favored the heterozygote or other forms of selection-maintaining diversity. In fact, either of these perspectives could readily be, and were, marshaled in support of eugenic perspectives that were common before and after the Second World War. For example, under the classical school, it was easy to postulate a genetic underclass that carried a greater-than-average load of deleterious mutations. Because natural selection pressures can be assumed to have differed among diverse human populations, it was possible to imagine that populations from some geographic regions would have a "superior" complement of variants in terms of key phenotypic characteristics as compared with other geographic regions.

Early "modern" approaches to quantifying biological differences were based on physical measurements that were heavily biased in the ways they were deployed. The distinguishing characteristics used to construct racial classification were those to which human perception is most finely tuned (skin color, eye shape and color, hair color and texture, etc.). Direct, objective methods of quantifying "genetic" variation (as opposed to "physical" characteristics) simply did not exist. Further, physical measurements were typically prone to environmental (nongenetic) influences, blurring the relationship between the measurement and genetic makeup of the individual.

The population characterization of the ABO blood group system by the Hirszfelds in the early 1900s, therefore, was seminal. It provided a biochemical marker that was closely aligned to underlying genetic variation, and, in so doing, provided the first major system for exploring patterns of human genetic diversity in an unbiased manner. Indeed, when tested in soldiers in armies of World War I, the pattern of A and B blood types showed frequency gradients that correlated with the geographic origin of the soldiers (Hirszfeld and Hirszfeld 1919). Soldiers from Western Europe (English and French) had a lower frequency of the "B" blood group, which appeared to gradually increase as one moved east toward Eastern European (Greeks, Turks, Russians) and Asian groups, suggesting that gene frequencies changed gradually across geographically defined populations.

A decisive break with the tradition of assuming sharp genetic divisions among ethnic groups came with the work of Richard Lewontin in 1972. Using multiple different polymorphic genetic markers, comprising blood group systems and serum protein markers that were ascertained in an unbiased manner, Lewontin generated data from more than 100 populations sampled across seven socially constructed "racial" groups (Caucasians, Africans, East Asians, South Asians, Amerindians, Oceanians, and Australians). Lewontin showed that the vast majority of human genetic diversity (~85%) is caused by individual differences that are shared across "all" populations and races. Only a small percentage (~15%) was because of differences "between" populations and a smaller percentage, again (~6%), was caused by differences between "racial" groups (Lewontin 1972). Lewontin's data suggested that although there is substantial genetic variation within the human population, such variation has accumulated over time; most of this variation appeared before the expansion of *Homo sapiens* out of Africa and the resulting isolation of populations within continents. Put another way, the considerable genetic differences we see between individuals has very little to do with so-called "racial" boundaries. Rather, it is merely the variation that was present in the original human population that seeded all the current human populations. Therefore, the same polymorphic alleles (genetic variants) are found in most populations, although their frequencies may differ substantially. Broadly speaking, there has been too little time for the accumulation of substantive divergence in a young species such as ours. The fact that the classical model predicted extensive genetic differentiation between populations was explained by the molecular evolution

pioneer Motoo Kimura, who hypothesized that most variation was selectively neutral (neither enhancing nor retarding human survival) and, therefore, largely free from the influence of Darwinian selective forces.

At the time of publication, Lewontin's findings were controversial, but consensus gradually emerged that genetic differences among populations are modest (Nei and Roychoudhury 1972; Cavalli-Sforza et al. 1994). Before Lewontin, the general consensus was that genetic diversity would be structured according to racial labels and, thus, the labels were scientifically justified. The observation that patterns in human genetic variation were largely gradual according to geographic boundaries and not subject to sudden population-specific changes that followed preconceived racial notions removed the biological argument for race (or, we would argue, it should have).

Lewontin illustrated that genetic variation was extensive and largely shared across populations. But, it was not until the sequencing of the first human genome (actually, a consensus of several human genomes) in 2003 that we appreciated just how extensive genetic variation really was in the human genome. Any two randomly selected individuals of European descent will differ at ~3 million points in their genome, or ~0.1% of their >3 billion bases of DNA. The fact that most of this variation is, in effect, selectively neutral presents an enormous challenge for characterizing those alleles that contribute to our common diseases in substantive ways. In other words, the challenge is to identify the few trait-altering variants that lie in an ocean of irrelevant ones.

A major breakthrough in this challenge was the development of GWAS. The basic framework used in these studies is to select key variants that inform about virtually all common variations in the human genome. These specially selected variants are often called tagging single-nucleotide polymorphisms (SNPs) because they are near perfect surrogates for variants not directly assayed. These variants could be tested easily by newly developed technologies using specially designed genotype chips. GWAS chips are also relatively inexpensive; one can now genotype a million variants for $<$\$50 a sample. Applied to large studies involving thousands of disease (case) and nondisease (control) individuals, the GWAS approach provided the framework to associate specific genetic variants and their cognate genomic regions with diseases, even if the study design was not well suited to identifying the actual genetically causal variants. The GWAS approach was successful in that it provided much needed momentum in the push to identify disease genes. Nevertheless, in most cases, even when applied to studies involving hundreds of thousands of participants, the approach failed to explain the majority of the presumed genetic component of any given trait (Manolio et al. 2009).

One explanation for this problem of "missing heritability" lies in the fact that the GWAS approach only tests for genetic variants that are common in a population, that is to say, those that Lewontin first observed as shared across individuals and populations. The reason for this is that the research community (through the International HapMap Project) had a good understanding of the nature and extent of common variation; it was, after all, "common" and therefore easy to find and test in large populations. Thus, it was a logical starting point for genome-wide studies. Further, it was not until the development of novel DNA-sequencing techniques in the last few years that the study of rare variants became logistically and financially feasible (Cirulli and Goldstein 2010). As a result, geneticists using GWAS in the late 2000s were akin to the drunken man who would only look for his lost keys under the streetlamp; he looked there because that is where the light was.

man groups. Perhaps the best evidence of this comes from the successor to the HapMap Project, the 1000 Genomes Project (The 1000 Genomes Project Consortium 2010), whose goal is to sequence the genomes of a large number of humans to provide a comprehensive survey of human genetic variation (Via et al. 2010). Investigators in the 1000 Genomes Project discovered that 63% of novel variants (that is, those that have never before been observed in humans) are found in African ancestry populations as compared with 33% with European ancestry.

In the same study, several hundred thousand SNPs with large allele-frequency differences were found across geographically distinct populations. Within these variants, there was enrichment for so-called "nonsynonymous" variants, which are characterized by important changes in the DNA sequence that lead to structural and functional changes in the proteins produced by these genes. This observation suggests that local populations adapted to their specific environments and the genetic changes that allowed this to happen were selected for by evolution (The 1000 Genomes Project Consortium 2010). These results also illustrate the fact that Lewontin's assessment related specifically to common variants because those are the ones most important to overall variation present in an individual. If you look at one individual, most of the variants that individual has are common variants, and those are the ones that follow Lewontin's pattern; they are mostly derived from the common human ancestral population. But, if the variants that are most important to phenotype variation are more rare, then this assessment that Lewontin provides does not apply to those most responsible for phenotypic variation.

THE PATTERN OF GEOGRAPHIC VARIATION FOR COMMON VARIATION MAY BE QUITE DIFFERENT FROM THAT OF VARIANTS INFLUENCING DISEASE RISK AND DRUG RESPONSE

Although most common variants are indeed common among most human populations, it has long been known that rarer gene variants can show markedly different patterns across hu-

MENDELIAN MUTATIONS ARE HIGHLY POPULATION SPECIFIC FOR A NUMBER OF REASONS

Because the Moravian monk Gregor Mendel was the first one to work out the basic laws of heredity, we refer to diseases within a family that "obviously" follow the rules of inheritance described by Mendel as Mendelian diseases. Typically, these mutations have a major effect on disease risk (and gene function) and relatively few genes can carry mutations that cause a given disease and still allow the organism to survive. Some of the mutations responsible for Mendelian diseases have long been known to show a high degree of population specificity. In some exceptional cases, this is clearly because of positive, as opposed to negative, natural selection. The autosomal recessive disease sickle cell anemia (that is caused by two defective copies of the β hemoglobin gene and, thus, producing a hemoglobin protein with reduced function), for example, is largely restricted to African, Mediterranean, and South Asian ancestry populations. In African-Americans, the allele frequency of the sickle hemoglobin (Hb S) mutation is ∼4% (Ashley-Koch et al. 2000). Why? Because although carrying two mutant Hb S alleles causes the devastating condition sickle cell anemia, carrying a single copy of Hb S does not usually cause health problems. However, it does protect against malarial infection. For this reason, it has been selected for in regions of the world where malaria has been endemic: Africa, the Mediterranean, and South Asia. Its frequency is significantly higher among populations that originate from these regions. In other words, carriers of one defective copy of the hemoglobin gene are at an evolutionary advantage in regions of the world where malaria is common and, therefore, this version of the hemoglobin gene has become more common in those areas (Aidoo et al. 2002).

Carrier advantage is not the principal reason why many Mendelian mutations can be thought of as more or less population specific. Most fundamentally, mutations that have a major impact on risk are rare because of natural selection against them. That means that they have been relatively recently introduced into the population by mutation, and the specific mutations are, therefore, usually geographically quite restricted. This, by itself, would mean that the mutations tend to be very different in different geographic regions, but not the total burden of diseases they cause. In fact, the collective frequency of disease-causing mutations in specific populations at specific genes can be quite

different from global averages because small population size and demographic history can also be important.

Consider the Ashkenazi Jews, who are statistically more likely to carry mutations that cause autosomal recessive Tay–Sachs disease in which affected children die at an early age because their mutations deprive them of a particular enzyme. In all likelihood, mutations causing Tay–Sachs increased in frequency during a time when the Ashkenazi Jewish population was small. Perhaps when the Ashkenazim were beginning to establish themselves in Europe during the early Middle Ages, one or more Tay–Sachs mutations arose by chance and the small breeding population led to a "founder effect," that is, persistence of particular alleles because those alleles were overrepresented when the population in question first emerged. Similarly, perhaps Tay–Sachs mutations were overrepresented after the Ashkenazi Jewish population underwent a "population bottleneck," that is, experienced a sharp contraction. For example, the European Jewish population declined precipitously following persecution of Jews during the First Crusade in the late 11th century and the subsequent spread of Black Death in the mid-14th century. It may well be that, by chance, Tay–Sachs mutations were present in surviving members of the Ashkenazim following these events and those mutations were, therefore, preferentially transmitted to subsequent generations (Slatkin 2004). Ashkenazi Jews' historical propensity to preferentially choose mates within their group has also served to keep Tay–Sachs alleles within their community at a relatively high frequency. Today, the carrier frequency of Tay–Sachs disease is on the order of 1 in 30 in self-identified Ashkenazi Jews, 10 times higher than in other populations. Before widespread population carrier screening of this disorder, 1 in 3600 children born to Ashkenazi parents had Tay–Sachs disease (Fernandes Filho and Shapiro 2004; Bray et al. 2010). Screening has since reduced Tay–Sachs births among Ashkenazim by some 90% (Ostrer and Skorecki 2013).

In other cases in which specific genetic diseases appear to be more common in a certain population, it is not clear whether the high frequency of rare disease-causing mutations is caused by chance, selective mating among carriers within the population, carrier selection advantage, or some combination of these factors. Cystic fibrosis, for example, is most common in European ancestry populations. In Caucasians, the frequencies of cystic fibrosis mutations in the cystic fibrosis transmembrane conductance regulator (CFTR) gene are significantly higher than in other populations and cause the autosomal recessive disease in 1 of 2500 newborns (Ratjen and Doring 2003). Over the years, geneticists have speculated as to why this is the case, often focusing on Darwinian selection as an explanation: perhaps carriers of CFTR mutations were more resistant to cholera and other dehydrating intestinal diseases (Bertranpetit and Calafell 1996). Or perhaps they were more resistant to contracting tuberculosis (Poolman and Galvani 2007). Another hypothesis suggested that carrier frequencies rose in Europe after farmers on the continent began raising dairy cattle, which led to the transmission of various pathogens from livestock to humans, perhaps via cow's milk (Alfonso-Sanchez et al. 2010). Although these ideas are intriguing, none have been proven to the extent of the implication of malaria as the selective force accounting for the rise of the Hb S allele. A particularly provocative hypothesis was promulgated by Harpending and Cochran that some of the mutations causing Tay–Sachs and other lysosomal storage diseases, several of which also occur at increased frequency in Ashkenazi Jews, were the result of positive selection; the idea was that somehow being a carrier for these diseases was associated with greater intelligence (Cochran et al. 2006). However, there remains no real evidence to support this speculation.

Whatever the reason for the emergence of disease-causing alleles at relatively high frequencies in specific populations, their existence suggests the possibility that rarer variants are also important in common diseases, and there may be more population specificity than anticipated by Lewontin's analysis of common variation. Consequently, we would do well to pay attention to the population frequencies of various human diseases and traits to better understand their genetic underpinnings.

COMMON VARIATION INFLUENCING DISEASE RISK AND DRUG RESPONSE

Even among common variants, some show relatively greater differentiation (frequency differences) among population groups because of genetic drift or selection, with clinically important consequences even at the level of the population average. One of the most well-known diseases for which common genetic variation affects both the spontaneous clearance of an infectious agent and treatment response is hepatitis C virus (HCV) infection. Treatment response refers to medical treatment with the combination of peginterferon-α (PegIFN-α) and antiviral therapies to induce viral clearance, whereas spontaneous clearance is the automatic viral clearance without exogenous drug administration. It was already well known that African ancestry individuals respond more poorly to HCV drug treatment than Caucasian and Asian individuals. In 2009, GWAS discovered a SNP (also known as rs12979860) in the *IL28B* locus (abbreviated as *IL28B* polymorphism below) that is highly associated with patient drug responses to medicines designed to treat HCV (Ge et al. 2009). Allele frequencies of *IL28B* polymorphism were found to differ largely among these ethnic populations, and explain the differences of treatment success rate among those populations. *IL28B* encodes interferon-λ-3, which is an important cytokine for innate immunity and one of the first responders to the invasion of foreign pathogens. Some believe the allele frequency of *IL28B* has been selected among different populations by one or more pathogen and, thus, evolved at different stages of human history. However, the exact natural selection pressure that causes the distinct pattern of allele frequency is unclear. Overall, the discovery of *IL28B* polymorphism illustrates that the frequency distribution of certain risk alleles is sufficient to affect the disease progression and drug responses. Below, we will discuss in detail how this common variant was discovered and its impact on both treatment-induced and spontaneous HCV clearance.

IL28B DISCOVERY FOR HCV TREATMENT RESPONSES

HCV is a positive-strand RNA virus belonging to the family *Flaviviridae*. HCV transmission is mainly through blood-to-blood contact and chronic infection usually results in fibrosis, cirrhosis, liver carcinoma, and even liver failure. It is estimated that 170 million people are chronically infected by HCV worldwide, and it is the major cause for liver transplants in the United States. Because HCV has been a serious public health problem in the United States and worldwide, there have been efforts to develop treatments for chronic HCV infection. However, the treatment success rate has been unsatisfactory. PegINF-α combined with ribavirin (RBV) therapy has been widely used to treat chronically infected HCV patients since 2002. The treatment success rate is moderate (from 20% to 70%) and is dependent on a patient's ancestry. Treatment success is defined as reaching sustained virological response (SVR), when the blood viral load is suppressed below the detectable level for

24 wk after 48 wk of combination treatment (Ghany et al. 2009). In East Asian populations, the PegIFN-α plus RBV treatment for chronically infected HCV patients has been shown to reach 76% of the overall SVR rate, which is dramatically higher than the 56% SVR rate of European-Americans and 24% of African-Americans (Liu et al. 2008; Ge et al. 2009). Before the genetic discovery of *IL28B*, the reason for the differences observed among major ethnic groups was unclear, and race had been used as a profiling feature to predict HCV treatment response.

The GWAS performed by Ge and colleagues, as well as studies performed by two other groups, identified a SNP (rs12979860) on the *IL28B* locus associated with the response of PegIFN-α plus RBV therapy. This genetic variant (rs12979860) is a C-to-T substitution with C being the major allele in Europeans and East Asians. The relative risk for SVR (chance to reach treatment success) is around threefold higher in C/C than non-C/C patients (including C/T and T/T), and is statistically highly significant (Fig. 1). Similar results were also found in several other studies; patients with the homozygous C/C genotype at *IL28B* generally have a two to three times higher treatment

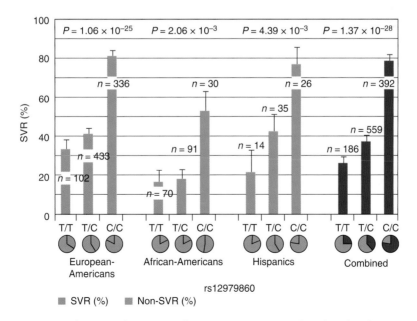

Figure 1. The SVR% as a function of ancestry and *IL28B* genotype. Figure based on data from Ge et al. (2009).

success rate than patients with C/T or T/T genotypes (Ge et al. 2009; Suppiah et al. 2009; Tanaka et al. 2009). European-American patients with C/C genotypes under different treatment regimens show ~80% SVR, compared with 30% and 40% SVR rates of C/T and T/T genotypes, respectively. In African-Americans, patients with the C/C genotype show ~50% of the SVR rate compared with <20% of the SVR rate for C/T and T/T patients (Fig. 1). The overall effect of the *IL28B* polymorphism is, therefore, substantial in predicting HCV treatment response. In general, regardless of ethnicity, the C/C genotype has higher SVR rate than non-C/C genotypes (twofold higher in European-Americans and Hispanics, and three-fold in African-Americans). This result suggests that C/C universally favors treatment success versus non-C/C, although in African-Americans, the same C/C genotype shows a lower SVR rate than in European-Americans (50% in African-Americans vs. 80% in European-Americans). The factors that cause this success rate difference in C/C genotype among individuals of different ethnicities are still unclear.

IL28B has received a great deal of attention since the GWAS discovery for its ability to predict the pretreatment drug response outcome, and the potential for its biological antiviral activities. Before the GWAS, the reason behind the link between ethnicity and drug responses was elusive, but we now clearly know that the *IL28B* allele frequencies show very different distributions across populations. Using random controls with unknown hepatitis C status, 90% of the East Asian population carried the *IL28B* C allele versus 70% in European-Americans. However, in the African-American population, the C allele has become the minor allele (smaller allele frequency) at 40%. Strikingly, according to the study performed by Ge and colleagues (2009), the C allele frequency showed linear correlation with the SVR rate in four distinct populations (Table 1). This concordance strongly suggests that the difference observed in HCV treatment response can be mostly explained by the allele frequency distribution among populations. In a subsequent study by Thomas et al. (2009), 51 geographical subpopulations were examined

Table 1. Correlation between SVR rate and *IL28B* C allele frequency

	SVR%	*IL28B* C allele frequency	Sample size
African-Americans	24%	0.40	191
Hispanics	51%	0.58	75
European-Americans	56%	0.63	871
East Asians	76%	0.95	154

Linear regression $r^2 = 0.93$. Data adapted from Ge et al. (2009).

for the *IL28B* polymorphism. The results were similar: the C allele frequency was highest in Asian populations, modest in European populations, and the lowest in African ancestry populations. This result showed the *IL28B* allele frequency distribution in higher resolution and corroborated the initial observations.

This correlation substantially explains the reason why different populations have significantly different treatment success rates. Up until now, the only gene for which there is strong evidence of an influence on HCV treatment response has been *IL28B*. An extensive search for other genetic factors that might contribute to HCV treatment response has been performed, but no statistically significant result for other genes that modify the effect of *IL28B* has been found thus far.

The profile based on race to predict treatment success rate in the past is now proven to be overly simplified. It is actually the *IL28B* genotype that plays a major role in determining treatment response, not ethnicity, and the differences observed among ethnicity can be explained merely by the allele frequency differences among geographic populations. HCV treatment response is a great example of how allele frequency can affect treatment outcomes among populations, and it seems highly likely that there will be other examples like this to be found in the future.

VARIATION OF *IL28B* ALSO AFFECTS SPONTANEOUS CLEARANCE OF HCV

Spontaneous clearance is the clearance of virus by the immune system without the administration of additional drugs. Based on studies of the natural history of HCV, 20%–30% of infected patients can spontaneously clear the virus, whereas the other 70%–80% become chronically infected and require drug therapy. The spontaneous clearance rate was estimated to be 36% in patients of non-African ancestry and 9% in patients of African ancestry (Thomas et al. 2000). Soon after the discovery of genetic association with treatment response for HCV, *IL28B* again was shown to be associated with the spontaneous clearance of HCV. Thomas and colleagues examined the *IL28B* polymorphism in six independent patient cohorts with the diagnosis of HCV infection. Patients were categorized as being chronically infected or having spontaneously cleared HCV by at least two blood tests separated by an interval of at least 6 months. Strikingly, the C allele of *IL28B* (rs12979860) also favors HCV clearance in these cohorts consisting of both European- and African-Americans. Individuals with the *IL28B* C/C genotype were, once again, two to three times more likely to clear the virus than the non-C/C patients. This result was similar to what had been observed in drug-induced HCV clearance. This finding suggests that *IL28B* has a universal effect on HCV resolution in natural settings without the administration of drugs, an important biological clue.

Because there is clear evidence of *IL28B* association with both treatment-induced and spontaneous viral clearance, it would be intriguing to know whether *IL28B* is also associated with the geographic distribution of HCV prevalence. However, the prevalence of HCV in major continents and *IL28B* frequency do not seem to be highly correlated. Although in most African countries, the prevalence rates are >3% of the total population (>3% is considered high prevalence for HCV), many East Asian countries (Asian populations have the highest rate of protective *IL28B* C allele) also comprise the majority of HCV chronic infections worldwide. For example, there is a 3.2% seroprevalence rate in China, which accounts for a major global HCV-infected population (Shepard et al. 2005). Many believe that country-specific features of the health care system itself may play

a major role in determining the likelihood of HCV exposure. For example, the availability of safe injections dramatically decreases the chance of exposure. Nevertheless, because HCV was first discovered in 1989, it has been impossible to obtain actual data of global HCV prevalence before industrialization.

Thanks to advances in tools for human genetic research, GWAS methods provide us with insight on common variants and infectious disease. In many carefully controlled clinical trials for HCV treatments, clear and consistent correlation between treatment success and the presence of the *IL28B* polymorphism has been shown. The genetic discovery shatters the long-lasting myth that race plays a role in HCV clearance. In fact, most of the difference in SVR rate can be explained solely by the frequency differences of *IL28B* alleles among populations. HCV infection, therefore, provides a salutary example of how common variation affects disease susceptibility and drug response.

IMPLICATIONS FOR HOW TO THINK ABOUT DISCOVERY AND ITS CLINICAL USE

The relative importance of rare and common variants in the traits that impose the greatest public health burden in the developed world remains unclear. Many believe that most cases of common disease are influenced by variants distributed across many genes, each with small effect, interacting with the environment in ways we do not yet understand. But we also know that sometimes cases of relatively common and certainly complex diseases can be caused by rare genetic changes of large effect. Among the best examples of the latter is neuropsychiatric disease, including conditions such as autism, epilepsy, and schizophrenia, in which rare, large genetic rearrangements (so-called "copy-number" variants) collectively account for a small but significant fraction of cases (Murdoch and State 2013). Another illustrative example of rare variants with large effects is epilepsy. In a recent report on two classical epileptic encephalopathies (infantile spasms and Lennox–Gastaut syndrome), researchers have discovered statistically significant enrichment of de novo mutations, that is, new variants that arise in the germline of the patient's parents, in specific gene sets. Some of these genes have significantly more de novo mutations in the patient cohort than would be expected by chance. This finding demonstrates that de novo mutations (occurring at one of several different genes) can have a strong influence on the risk of epilepsy (Epi4K Consortium, Epilepsy Phenome/Genome Project 2013).

A related question concerns the proportion of the functional genetic variation that is present in the human population as a result of some form of carrier advantage, as is clearly important in sickle cell anemia, or some sort of mutation-selection balance (in which the arisal of de novo mutations is balanced by their loss because they reduce the fitness of the mutation bearer) as is clearly responsible for the "copy number" variants mentioned above. In the latter, genetic rearrangements can lead to changes in human cognitive potential, but in ways that we cannot yet predict with a high degree of confidence.

Despite the clear similarity of human populations as described by Lewontin, the few examples listed here show that no matter what sort of evolutionary tradeoffs existed in the past of the human species, the genetic bases of medically relevant traits can be profoundly different at both the individual and population levels. The nature of these differences will obviously depend, in large part, on what sort of genetic variation causes most diseases. More generally, we still do not exactly know whether the majority of important variants is generally deleterious and present because of mutation-selection balance, or whether the important ones are more nuanced in their effects in that they are sometimes helpful, sometimes harmful.

It is clear that geneticists cannot assume that aspects of human genetic disease and other medically relevant traits can be understood by studying only one population because human groups are neither homogeneous nor genetic studies fruitful unless they are population comparative. As such, research programs that concentrate their discovery efforts in populations of European ancestry alone, as most genomics ef-

forts have performed to date (Need and Goldstein 2009; Bustamante et al. 2011), are inefficient and incomplete. If we are to fulfill the promise of the Human Genome Project, enhance biological discovery, and begin to bring our knowledge of population genetics to bear on long-standing health disparities, then we must understand and appreciate the enormous range of variation within our species. As the poet Audra Lorde wrote, "It is not our differences that divide us. It is our inability to recognize, accept, and celebrate those differences."

REFERENCES

Aidoo M, Terlouw DJ, Kolczak MS, McElroy PD, ter Kuile FO, Kariuki S, Nahlen BL, Lal AA, Udhayakumar V. 2002. Protective effects of the sickle cell gene against malaria morbidity and mortality. *Lancet* **359:** 1311–1312.

Alfonso-Sanchez MA, Perez-Miranda AM, Garcia-Obregon S, Pena JA. 2010. An evolutionary approach to the high frequency of the ΔF508 CFTR mutation in European populations. *Med Hypotheses* **74:** 989–992.

Ashley-Koch A, Yang Q, Olney RS. 2000. Sickle hemoglobin (HbS) allele and sickle cell disease: A HuGE review. *Am J Epidemiol* **151:** 839–845.

Bertranpetit J, Calafell F. 1996. Genetic and geographical variability in cystic fibrosis: Evolutionary considerations. *Ciba Found Symp* **197:** 97–114; discussion 114–118.

Bray SM, Mulle JG, Dodd AF, Pulver AE, Wooding S, Warren ST. 2010. Signatures of founder effects, admixture, and selection in the Ashkenazi Jewish population. *Proc Natl Acad Sci* **107:** 16222–16227.

Bustamante CD, Burchard EG, De la Vega FM. 2011. Genomics for the world. *Nature* **475:** 163–165.

Cavalli-Sforza LL, Menozzi P, Piazza A. 1994. *The history and geography of human genes.* Princeton University Press, Princeton, NJ.

Cirulli ET, Goldstein DB. 2010. Uncovering the roles of rare variants in common disease through whole-genome sequencing. *Nat Rev Genets* **11:** 415–425.

Cochran G, Hardy J, Harpending H. 2006. Natural history of Ashkenazi intelligence. *J Biosoc Sci* **38:** 659–693.

Epi4K Consortium, Epilepsy Phenome/Genome Project. 2013. De novo mutations in epileptic encephalopathies. *Nature* **501:** 217–221.

Fernandes Filho JA, Shapiro BE. 2004. Tay–Sachs disease. *Arch Neurol* **61:** 1466–1468.

Ge D, Fellay J, Thompson AJ, Simon JS, Shianna KV, Urban TJ, Heinzen EL, Qiu P, Bertelsen AH, Muir AJ, et al. 2009. Genetic variation in IL28B predicts hepatitis C treatment-induced viral clearance. *Nature* **461:** 399–401.

Ghany MG, Strader DB, Thomas DL, Seeff LB, American Association for the Study of Liver Diseases. 2009. Diagnosis, management, and treatment of hepatitis C: An update. *Hepatology* **49:** 1335–1374.

Hirszfeld L, Hirszfeld H. 1919. Essai d'application des methodes au problem des races [Testing the application of methods to the question of race]. *Anthropologie* **29:** 505–537.

Lewontin R. 1972. The apportionment of human diversity. In *Evolutionary biology* (ed. Dobzhansky T, Hecht M, Steere W), pp. 381–398. Appleton Centuary Crofts, New York.

Liu CH, Liu CJ, Lin CL, Liang CC, Hsu SJ, Yang SS, Hsu CS, Tseng TC, Wang CC, Lai MY, et al. 2008. Pegylated interferon-α-2a plus ribavirin for treatment-naive Asian patients with hepatitis C virus genotype 1 infection: A multicenter, randomized controlled trial. *Clin Infect Dis* **47:** 1260–1269.

Lorde A. 2007. *Sister outsider: Essays and speeches.* Ten Speed Press, New York.

Manolio TA, Collins FS, Cox NJ, Goldstein DB, Hindorff LA, Hunter DJ, McCarthy MI, Ramos EM, Cardon LR, Chakravarti A, et al. 2009. Finding the missing heritability of complex diseases. *Nature* **461:** 747–753.

Murdoch JD, State MW. 2013. Recent developments in the genetics of autism spectrum disorders. *Curr Opin Genet Dev* **23:** 310–315.

Need AC, Goldstein DB. 2009. Next generation disparities in human genomics: Concerns and remedies. *Trends Genet* **25:** 489–494.

Nei M, Roychoudhury AK. 1972. Gene differences between Caucasian, Negro, and Japanese populations. *Science* **177:** 434–436.

Ostrer H, Skorecki K. 2013. The population genetics of the Jewish people. *Hum Genet* **132:** 119–127.

Poolman EM, Galvani AP. 2007. Evaluating candidate agents of selective pressure for cystic fibrosis. *J R Soc Interface* **4:** 91–98.

Ratjen F, Doring G. 2003. Cystic fibrosis. *Lancet* **361:** 681–689.

Shepard CW, Finelli L, Alter MJ. 2005. Global epidemiology of hepatitis C virus infection. *Lancet Infect Dis* **5:** 558–567.

Slatkin M. 2004. A population-genetic test of founder effects and implications for Ashkenazi Jewish diseases. *Am J Hum Genet* **75:** 282–293.

Suppiah V, Moldovan M, Ahlenstiel G, Berg T, Weltman M, Abate ML, Bassendine M, Spengler U, Dore GJ, Powell E, et al. 2009. IL28B is associated with response to chronic hepatitis C interferon-α and ribavirin therapy. *Nat Genet* **41:** 1100–1104.

Tanaka Y, Nishida N, Sugiyama M, Kurosaki M, Matsuura K, Sakamoto N, Nakagawa M, Korenaga M, Hino K, Hige S, et al. 2009. Genome-wide association of IL28 with response to pegylated interferon-α and ribavirin therapy for chronic hepatitis C. *Nat Genet* **41:** 1105–1109.

The 1000 Genomes Project Consortium. 2010. A map of human genome variation from population-scale sequencing. *Nature* **467:** 1061–1073.

Thomas DL, Astemborski J, Rai RM, Anania FA, Schaeffer M, Galai N, Nolt K, Nelson KE, Strathdee SA, Johnson L, et al. 2000. The natural history of hepatitis C virus infection: Host, viral, and environmental factors. *JAMA* **284:** 450–456.

Thomas DL, Thio CL, Martin MP, Qi Y, Ge D, O'Huigin C, Kidd J, Kidd K, Khakoo SI, Alexander G, et al. 2009. Genetic variation in IL28B and spontaneous clearance of hepatitis C virus. *Nature* **461:** 798–801.

Torkamani A, Pham P, Libiger O, Bansal V, Zhang G, Scott-Van Zeeland AA, Tewhey R, Topol EJ, Schork NJ. 2012. Clinical implications of human population differences in genome-wide rates of functional genotypes. *Front Genet* **3:** 211.

Via M, Gignoux C, Burchard EG. 2010. The 1000 Genomes Project: New opportunities for research and social challenges. *Genome Med* **2:** 3.

Index